Studies in Economic Theory

Springer
Berlin
Heidelberg
New York
Barcelona
Hong Kong
London
Milan
Paris
Singapore
Tokyo

Titles in the Series

M. Ali Khan and Nicholas C. Yannelis (Eds.)
Equilibrium Theory in Infinite Dimensional Spaces

Charalambos D. Aliprantis, Kim C. Border and
Wilhelmus A. J. Luxemburg (Eds.)
Positive Operators, Riesz Spaces, and Economics

Donald G. Saari
Geometry of Voting

Charalambos D. Aliprantis and Kim C. Border
Infinite Dimensional Analysis

Jean-Pierre Aubin
Dynamic Economic Theory

Mordecai Kurz (Ed.)
Endogenous Economic Fluctuations

Jean-François Laslier
Tournament Solutions and Majority Voting

Ahmet Alkan, Charalambos D. Aliprantis and
Nicholas C. Yannelis (Eds.)
Theory and Applications

James C. Moore
Mathematical Methods for Economic Theory 1

James C. Moore
Mathematical Methods for Economic Theory 2

Mukul Majumdar, Tapan Mitra and Kazuo Nishimura
Optimization and Chaos

Katrin Sieberg
Criminal Dilemmas

Monique Florenzano and Cuong Le Van
Finite Dimensional Convexity and Optimization

Monique Florenzano · Cuong Le Van

In cooperation with Pascal Gourdel

Finite Dimensional Convexity and Optimization

With 9 Figures

Springer

Monique Florenzano
Cuong Le Van
Pascal Gourdel
MSE-Université de Paris 1
CERMSEM
106–112 boulevard de l'Hopital
75013 Paris
France

ISBN 3-540-41516-5 Springer-Verlag Berlin Heidelberg New York

Library of Congress Cataloging-in-Publication Data
Die Deutsche Bibliothek - CIP-Einheitsaufnahme
Florenzano, Monique: Finite Dimensional Convexity and Optimization / Monique Florenzano; Cuong Le Van. In cooperation with Pascal Gourdel. - Berlin; Heidelberg; New York; Barcelona; Hong Kong; London; Milan; Paris; Singapore; Tokyo: Springer, 2001
(Studies in economic theory; 13)
ISBN 3-540-41516-5

Springer-Verlag Berlin Heidelberg New York
a member of BertelsmannSpringer Science+Business Media GmbH

Printed in Germany

Hardcover Design: Erich Kirchner, Heidelberg

SPIN 10794588 42/2202-5 4 3 2 1 0 – Printed on acid-free paper

Preface

Interest in convexity arises from areas of application concerned with fixed point theory and optimization. The primary aim of this book is to present notions of convex analysis which constitute the basic underlying structure of argumentation in Economic Theory and which are common to optimization problems encountered in many applications.

The topological framework is that of a finite dimensional vector space endowed with the Euclidean norm. The first three chapters are devoted to the notions and main results of convexity in $\mathbb{R}^n$. Most of these notions and results could be formulated in a broader setting. We deliberately limited ourselves to the finite dimensional case, but as a counter-part, we use extensively the specific characteristics of the Euclidean space in the proofs. Linear programming, which is the topic of Chapter 4, has numerous applications, in particular in operations research.

The analysis of convex functions starts only with Chapter 5. However, the duality of this approach should not conceal the link between convex functions and sets. Any convex function may be associated with its epigraph - a convex set. Conversely, many convex sets may be studied by means of the convex function of which they are the epigraph. Therefore, the reader has to go back and forth between the first chapters and the second part of the book which starts with Chapter 5.

Convex programming, a generalization of linear programming, has also many applications. There again, going back to the first part, one may observe that the necessary and sufficient optimality conditions in linear programming are nothing else than the Kuhn and Tucker conditions. However, the duality in linear programming, which is treated in Chapter 4 by techniques resulting from Minkowski-Farkas' Lemma, goes further than a simple application of the convex duality to the linear case would allow for. The last chapter is devoted to nonconvex optimization under constraints. We continue by studying the case quasi-convex objective functions which is frequently encountered in micro-economics. Then, we treat differentiable objective functions (not necessarily quasi-convex or convex).

The intended readers of this book are senior undergraduate students, graduate students and specialists of mathematical programming whose research fields are applied mathematics and economics. It should be useful to any researcher desirous of an initiation into optimization techniques. The limitation to the finite dimensional case responds to the necessities of a diverse audience. The level of the book stays relatively elementary in the sense that the only prerequisites concern basic notions of linear algebra and of general topology. In this framework, our presentation does not make any concession in terms of the rigor of the proofs and the generality of the notions introduced. Exercises appear at the end of the various chapters, yielding complements and applications. Some are intended to be of a rather routine nature or to apply the details deliberately omitted in the text. Some others, however, give significant and useful additional results. They are taken from books cited in the references and from lecture notes provided by our colleagues.

The results contained in this book, although classic nowadays, are linked to the construction of Functional Analysis at the beginning of the XXth century, as well as to contributions in the Fifthties and later of G. Dantzig, W. Fenchel, D. Gale, V.L. Klee, H.W. Kuhn, T. Rockafellar, A.W. Tucker, H. Weyl and many others, in a period of revived activity in Convex analysis, in relation to new fields in Optimization, Economics and Game Theory. Whereas we do not discuss what is now a substantial literature dealing with nonsmooth analysis, or with convex analysis in its infinite dimensional setting, one of our subsidiary aims is to present the material in such a way it will aid and encourage the reader to proceed to these subjects. The list of references given at the end of the book is partial in both senses of the term. The readers consulting these references will notice that they go well beyond the scope of the present book. In particular, they will find there a more systematic definition of duality, extensions to infinite dimensions, a study of algorithms and an introduction to applications.

We are grateful to Roko Aliprantis for his invitation to contribute the book to this series and to Werner Müller, our editor at Springer-Verlag, for his support. We are also grateful to the Universities Paris I and Paris V for having given us the opportunity to teach mathematical programming at different levels. We thank our colleagues and friends whose lecture notes and exercise sheets we consulted : J. Abdou, A. Auslender, M. Balinski, J.M. Bonnisseau, B. Cornet, R.A. Dana, J.Y. Jaffray, R. Pallu de la Barrière, M. Théra. Special thanks also go to Nizar Allouch, Yasar N. Barut, Ali Khan for reading the manuscript and providing numerous corrections. Pascal Gourdel did not participate in the planning of this volume, but supervised its preparation in all of its subsequent stages and in its various respects. We are especially indebted to him for his contributions which exceeded all bounds of collegial help and assistance. We typed and typeset this text ourselves and are responsible for all remaining errors, mathematical or otherwise.

Paris
December 2000

Monique Florenzano
Cuong Le Van

Contents

Notation

When it is not basic, a precise definition for the terminology introduced below is to be found in the course of the book (with the help of the index).

Given a positive integer n,

$\mathbb{R}^n$:= n-dimensional Euclidean space;
$\mathbb{R}^n_+$:= the nonnegative orthant of $\mathbb{R}^n$;
$\mathbb{R} := \mathbb{R}^1$;
$\mathbb{R}_+ := \mathbb{R}^1_+$;
$\overline{\mathbb{R}} := \mathbb{R} \cup \{-\infty\} \cup \{+\infty\}$.

For a real sequence $(\lambda_\nu)_{\nu \in \mathbb{N}}$,

$\lambda_\nu \downarrow 0$ means that the sequence is decreasing with infimum 0.

For any $x,\ y \in \mathbb{R}^n$, and $r > 0$ in $\mathbb{R}$,

x_i:= the ith coordinate of x, $i = 1, \dots, n$;
$x \cdot y := \sum_{i=1}^n x_i y_i$ denotes the Euclidean inner product of x and y;
$\|x\| := \sqrt{x \cdot x}$ denotes the Euclidean norm of x;
$x \geq y$ means $x_i \geq y_i$ for every $i = 1, \dots, n$;
$x > y$ means $x \geq y$ and $x \neq y$;
$x \gg y$ means $x_i > y_i$ for every $i = 1, \dots, n$;
$x^+ := x \vee 0 = \left(\max\{x_i, 0\}\right)_{i=1}^n$ denotes the positive part of x;
$x^- := (-x) \vee 0 = \left(\max\{-x_i, 0\}\right)_{i=1}^n$ denotes the negative part of x;
$B(x,r) := \{z \in \mathbb{R}^n \mid \|z - x\| < r\}$ denotes the open ball with center x and radius r;
$\overline{B(x,r)} := \{z \in \mathbb{R}^n \mid \|z - x\| \leq r\}$ denotes the closed ball with center x and radius r;
$[x, y]$ denotes the closed line segment joining x and y;
$]x, y[$ or (x, y) denote the open line segment joining x and y.

For any subsets A, B of $\mathbb{R}^n$, for any $\lambda \in \mathbb{R}$,

$A + B := \{z \in \mathbb{R}^n \mid z = x + y,\ x \in A,\ y \in B\}$;
$A - B := \{z \in \mathbb{R}^n \mid z = x - y,\ x \in A,\ y \in B\}$;
$\lambda A := \{z \in \mathbb{R}^n \mid z = \lambda x,\ \lambda \in \mathbb{R},\ x \in A\}$.

The algebraic concepts $A + B$ and $A - B$ should not be confused with the set-theoretic concepts $A \cup B$ and $A \setminus B$.

For any subset A of $\mathbb{R}^n$,
$\operatorname{co} A$ denotes the convex hull of A;
$\operatorname{aff} A$ denotes the affine hull of A;
$\underline{\mathrm{K}}(A)$ denotes the convex cone generated by A;
$\overline{A}$ denotes the closure of A;
$\mathrm{i}(A)$ denotes the interior of A;
∂A denotes the boundary of A;
$\overline{\operatorname{co}}\, A$ denotes the closed convex hull of A;
$\dim A$ denotes the dimension of A;
A° is the polar of A.

For any convex subset C of $\mathbb{R}^n$,
$\mathrm{ri}(C)$ denotes the relative interior of A;
$\mathrm{Bd}^{\mathrm{r}}\, C := \overline{C} \setminus \mathrm{ri}(C)$ denotes the relative boundary of C;
$\mathrm{A}(C)$ denotes the asymptotic (or recession) cone of C;
$\mathrm{ext}(C)$ denotes the set of extreme points of C;
$\exp(C)$ denotes the set of exposed points of C;
$N_C(x)$ denotes the normal cone to C at $x \in C$;
$\delta_C : \mathbb{R}^n \to \mathbb{R}$ denotes the indicator function of C;
$\delta_C^* : \mathbb{R}^n \to \overline{\mathbb{R}}$ denotes the support function of C;
$\gamma_C : \mathbb{R}^n \to \overline{\mathbb{R}}$ denotes the gauge function of C.

If F is a linear subspace of $\mathbb{R}^n$, $F^\perp$ is the orthogonal complement of F.

For a real $m \times n$-matrix A,
${}^tA$ is the transpose of matrix A.
As usual, a vector x of $\mathbb{R}^n$ is identified with a $(n \times 1)$-matrix and ${}^tx$ is a $(1 \times n)$-matrix.

For a function $f : \mathbb{R}^n \to \overline{\mathbb{R}}$,
$\mathrm{dom}(f)$ denotes the domain of f;
$\mathrm{epi}(f)$ denotes the epigraph of f;
$\mathrm{lev}_{\leq\alpha}(f)$, $\mathrm{lev}_{<\alpha}(f)$, $\mathrm{lev}_{\geq\alpha}(f)$, $\mathrm{lev}_{>\alpha}(f)$ denote the level sets of f;
$f'(x;y)$ denotes the derivative of f at x in the direction of y;
$\partial f(x)$ denotes the subdifferential of f at x;
$\nabla f(x)$ denotes the gradient of f at x;
$\frac{\partial f}{\partial x_i}(x)$ denotes the ith partial derivative of f at x.

If f is defined on an interval of $\mathbb{R}$,
$f'_+(x) := f'(x;1)$;
$f'_-(x) := f'(x;-1)$.

1. Convexity in $\mathbb{R}^n$

1.1 Basic concepts

1.1.1 Line segment, convex set and convex hull

Let x and y be two points of $\mathbb{R}^n$. The set

$$[x, y] := \{z \in \mathbb{R}^n \mid z = \alpha x + (1-\alpha)y,\ \alpha \in \mathbb{R} \text{ and } 0 \le \alpha \le 1\}$$

is called *closed line segment* joining x and y. One also denotes

$$[x, y) := [x, y] \setminus \{y\} = \{z \in \mathbb{R}^n \mid z = \alpha x + (1-\alpha)y,\ \alpha \in \mathbb{R} \text{ and } 0 < \alpha \le 1\},$$

$$(x, y] := [x, y] \setminus \{x\} = \{z \in \mathbb{R}^n \mid z = \alpha x + (1-\alpha)y,\ \alpha \in \mathbb{R} \text{ and } 0 \le \alpha < 1\}.$$

The set

$$\begin{aligned}(x, y) :&= [x, y] \setminus \{x, y\} \\ &= \{z \in \mathbb{R}^n \mid z = \alpha x + (1-\alpha)y,\ \alpha \in \mathbb{R} \text{ and } 0 < \alpha < 1\}\end{aligned}$$

is called *open line segment* joining x and y. It may also be noted $]x, y[$.

A subset C of R^n is said to be *convex* if it contains every closed line segment joining two points of C. In other words, C is convex if and only if for x and y in C and $\alpha \in \mathbb{R}$ such that $0 \le \alpha \le 1$, $\alpha x + (1-\alpha)y \in C$ holds true. By induction on the integer p, it is easily seen that C is convex if and only if $\sum_{k=1}^{p} \lambda_k x^k \in C$ for every finite set $\{x^1, \ldots, x^p\}$ of p elements of C and for every system of p nonnegative real coefficients $\{\lambda_1, \ldots, \lambda_p\}$ such that $\sum_{k=1}^{p} \lambda_k = 1$. The quantity $\sum_{k=1}^{p} \lambda_k x^k$ is called *convex combination* of $\{x^1, \ldots, x^p\}$. Hence a subset C of R^n is convex if and only if every convex combination of finitely many elements of C belongs to C.

The empty set $\emptyset$, $\{a\}$ for any $a \in \mathbb{R}^n$ and $\mathbb{R}^n$ are convex sets. Any intersection of convex subsets of $\mathbb{R}^n$ is a convex subset of $\mathbb{R}^n$. With any arbitrary subset A of $\mathbb{R}^n$, we can associate a convex set, called the *convex hull* of A and denoted by $\operatorname{co} A$, which is the intersection of all convex subsets of $\mathbb{R}^n$ containing A. Obviously, it is the smallest convex subset of $\mathbb{R}^n$ containing A.

Proposition 1.1.1. *The convex hull of a subset A of $\mathbb{R}^n$ is precisely the set of all convex combinations of finitely many members of A.*

Proof. Let $x = \sum_{k=1}^{p} \lambda_k x^k$ with $\{x^1, \dots, x^p\} \subset A$, $\{\lambda_1, \dots, \lambda_p\} \subset \mathbb{R}_+$ and $\sum_{k=1}^{p} \lambda_k = 1$. The point x obviously belongs to any convex set containing A. Conversely, it is easily seen that the set $\{\sum_{k=1}^{p} \lambda_k x^k \mid p \in \mathbb{N},\ \{x^1, \dots, x^p\} \subset A,\ \{\lambda_1, \dots, \lambda_p\} \subset \mathbb{R}_+,\ \sum_{k=1}^{p} \lambda_k = 1\}$ is a convex set containing A, hence containing $\operatorname{co} A$. □

Proposition 1.1.1 holds true for an arbitrary real vector space. A sharper result in $\mathbb{R}^n$ is

Proposition 1.1.2 (Caratheodory). *The convex hull of A is precisely the set of all convex combinations of at most $(n+1)$ elements of A.*

Proof. Let $x \in \operatorname{co} A$. By the previous proposition, for some $p \in \mathbb{N}$, $x = \sum_{k=1}^{p} \lambda_k x^k$ with $x^k \in A$, $\lambda_k \geq 0$, $k = 1, \dots, p$ and $\sum_{k=1}^{p} \lambda_k = 1$. Without loss of generality, one can assume that $\lambda_k > 0$ for all $k = 1, \dots, p$.

If $p \leq n+1$, x has the required form. If $p > n+1$, the $(p-1)$ points $(x^2 - x^1), \dots, (x^p - x^1)$ are linearly dependent. Let $\alpha_2, \dots, \alpha_p \in \mathbb{R}$ not all equal to 0 be such that $\sum_{k=2}^{p} \alpha_k x^k - (\sum_{k=2}^{p} \alpha_k) x^1 = 0$. In other words, there exist $\beta_1, \dots, \beta_p \in \mathbb{R}$ not all equal to 0, such that $\sum_{k=1}^{p} \beta_k x^k = 0$ and $\sum_{k=1}^{p} \beta_k = 0$. Define

$$
\begin{aligned}
I &= \big\{k \in \{1, \dots, p\} \mid \beta_k < 0\big\} \\
J &= \big\{k \in \{1, \dots, p\} \mid \beta_k > 0\big\} \\
\theta &= \min_{k \in J} \frac{\lambda_k}{\beta_k} \\
K &= \{k \in J \mid \lambda_k - \theta\beta_k = 0\}.
\end{aligned}
$$

Since I and J are nonempty, $\theta > 0$ and K is nonempty. Then $x = \sum_{k=1}^{p} \lambda_k x^k = \sum_{k=1}^{p} (\lambda_k - \theta\beta_k) x^k = \sum_{k \notin K} (\lambda_k - \theta\beta_k) x^k$. Note that $k \in I$ implies $\lambda_k - \theta\beta_k > 0$; $k \in J$ implies $\lambda_k - \theta\beta_k \geq 0$; $k \in J \backslash K$ implies $\lambda_k - \theta\beta_k > 0$; $\sum_{k=1}^{p} (\lambda_k - \theta\beta_k) = \sum_{k=1}^{p} \lambda_k - \theta \sum_{k=1}^{p} \beta_k = 1$. Thus for representing x, a convex combination (with strictly positive coefficients) of p elements of A has been replaced by a convex combination (with strictly positive coefficients) of q elements of A, with $q < p$.

The procedure can be repeated until $q \leq n+1$ is obtained. □

1.1.2 Line, affine set and affine hull

Let x and y be two different points of $\mathbb{R}^n$. The set

$$\{z \in \mathbb{R}^n \mid z = x + \lambda(y - x),\ \lambda \in \mathbb{R}\}$$

is called *line* passing through x and y. The set

$$\{z \in \mathbb{R}^n \mid z = x + \lambda(y - x),\ \lambda \in \mathbb{R}_+\}$$

is called *half-line* emanating from x and passing through y. A subset V of $\mathbb{R}^n$ is called *affine set* if $x, y \in V$ imply that the whole line passing through x and y is included in V. In other words, V is an affine set if and only if $\lambda x + (1-\lambda)y \in V$ holds true, whatever be x and y in V and $\lambda \in \mathbb{R}$. As previously, it is easily seen, by induction on p, that V is an affine set if and only if $\sum_{k=1}^{p} \lambda_k x^k \in V$ for every finite set $\{x^1, \ldots, x^p\}$ of p elements of V and for every system of p real coefficients $\{\lambda_1, \ldots, \lambda_p\}$ such that $\sum_{k=1}^{p} \lambda_k = 1$. The quantity $\sum_{k=1}^{p} \lambda_k x^k$ is called *affine combination* of $\{x^1, \ldots, x^p\}$. Hence V is an affine set if and only if every affine combination of finitely many elements of V belongs to V.

The empty set $\emptyset$, $\{a\}$ for any $a \in \mathbb{R}^n$ and $\mathbb{R}^n$ are affine sets. Any intersection of affine sets of $\mathbb{R}^n$ is an affine set of $\mathbb{R}^n$. With any arbitrary subset A of $\mathbb{R}^n$, we can associate an affine set, called the *affine hull* of A and denoted by $\operatorname{aff} A$, which is the intersection of all affine sets of $\mathbb{R}^n$ containing A. The following proposition is an analogue of Proposition 1.1.1. Its proof is similar.

Proposition 1.1.3. *The affine hull of a subset A of $\mathbb{R}^n$ is precisely the set of all affine combinations of finitely many members of A.*

Proof. Let $x = \sum_{k=1}^{p} \lambda_k x^k$ with $\{x^1, \ldots, x^p\} \subset A$, $\{\lambda_1, \ldots, \lambda_p\} \subset \mathbb{R}$ and $\sum_{k=1}^{p} \lambda_k = 1$. The point x obviously belongs to any affine set containing A. Conversely, it is easily seen that the set $\{\sum_{k=1}^{p} \lambda_k x^k \mid p \in \mathbb{N},\ \{x^1, \ldots, x^k\} \subset A,\ \{\lambda_1, \ldots, \lambda_p\} \subset \mathbb{R},\ \sum_{k=1}^{p} \lambda_k = 1\}$ is an affine set containing A, hence containing $\operatorname{aff} A$. □

Recall that for two subsets A and B of $\mathbb{R}^n$, $A + B = \{x \in \mathbb{R}^n \mid x = a + b,\ a \in A,\ b \in B\}$. Two affine sets V and V' of $\mathbb{R}^n$ are said to be *parallel* if V' can be deduced from V by translation, that is, if there exits $a \in \mathbb{R}^n$ such that $V' = \{a\} + V$. Parallelism is an equivalence relation. The following proposition characterizes affine sets as being translates of some linear subspace of $\mathbb{R}^n$, unique for each affine set V and which defines the equivalence class of V. It allows for deducing the geometry of affine sets from the properties of linear subspaces of $\mathbb{R}^n$.

Proposition 1.1.4. *The linear subspaces of $\mathbb{R}^n$ are precisely the affine sets containing the origin. Moreover, a subset V of $\mathbb{R}^n$ is an affine set if and only if there exist a point a and a linear subspace M of $\mathbb{R}^n$ such that $V = \{a\} + M$. M is unique and equal to $V - V$; in counterpart, $V = \{x\} + M$ for every $x \in V$.*

Proof. A linear subspace is obviously an affine set containing 0. Conversely, recall that M is a linear subspace if and only if $\lambda, \mu \in \mathbb{R}$ and $x, y \in M$ imply $\lambda x + \mu y \in M$. The last relation is equivalent to $\lambda x + \mu y + (1 - \lambda - \mu)0 \in M$. Thus an affine set containing 0 is a linear subspace. This proves the first assertion.

If M is a linear subspace, $\{a\} + M$ is easily seen to be an affine set. Assume conversely that V is an affine set. Using as above convenient affine combinations, one verifies successively that $M = V - V$ is a linear subspace and that for any $x \in V$, $V = \{x\} + M$.

Finally, if $V = \{a\} + M$, one first remarks that $a \in V$. Then $M \subset V - V$. Since obviously $V - V \subset M$, the formula $M = V - V$ proves the uniqueness of M. □

The last assertion of Proposition 1.1.4 is a particular case of the following proposition:

Proposition 1.1.5. *Let A be a subset of $\mathbb{R}^n$. Then for any $a \in \operatorname{aff} A$, $\operatorname{aff} A = \{a\} + M$ where M is the linear hull (span) of $A - A$ (that is the linear subspace of $\mathbb{R}^n$ generated by $A - A$).*

Proof. Choose any $a \in \operatorname{aff} A$. Then, $\operatorname{aff} A - \operatorname{aff} A$ is a linear subspace of $\mathbb{R}^n$ which contains $A - A$, so that $M \subset \operatorname{aff} A - \operatorname{aff} A$ and $\{a\} + M \subset \operatorname{aff} A + \operatorname{aff} A - \operatorname{aff} A \subset \operatorname{aff} A$. Conversely, choose any $x \in A$. From the identity $x = a + (x - a)$, one deduces that $x \in \{a\} + M$ and thus $A \subset \{a\} + M$, which implies $\operatorname{aff} A \subset \{a\} + M$. □

1.1.3 Affine maps

Let V_1 and V_2 be two affine sets. A map $f : V_1 \to V_2$ is said to be *affine* if for each pair $x, y \in V_1$ and $\alpha \in \mathbb{R}$ one has

$$f\big(\alpha x + (1 - \alpha)y\big) = \alpha f(x) + (1 - \alpha) f(y).$$

One then has:

$$f(\sum_{k=1}^{p} \lambda_k x^k) = \sum_{k=1}^{p} \lambda_k f(x^k),$$

for every finite set $\{\lambda_1, \ldots, \lambda_p\}$ of p real numbers verifying $\sum_{k=1}^{p} \lambda_k = 1$. If A is a convex (resp. an affine) subset of V_1 , $f(A)$ is a convex (resp. an affine) subset of V_2. If A is any subset of V_1, $f(\operatorname{co} A) = \operatorname{co} f(A)$ (resp. $f(\operatorname{aff} A) = \operatorname{aff} f(A)$).

An affine map $f : V \to \mathbb{R}$ is said to be an *affine functional* on V.

Proposition 1.1.6. *The affine maps from $\mathbb{R}^n$ into $\mathbb{R}^m$ are precisely the maps defined by $f(x) = Ax + b$ for some $(m \times n)$-matrix A and a constant b such that $b = f(0)$.*

Proof. Let $f : \mathbb{R}^n \to \mathbb{R}^m$ be affine. Define g by $g(x) = f(x) - f(0)$. One can easily verify that g is linear. If A is the $(m \times n)$-matrix such that $g(x) = Ax$, one has $f(x) = Ax + f(0)$. If, conversely, $f(x) = Ax + b$, one verifies that f is affine and that $b = f(0)$. □

An immediate consequence of the previous proposition is that every affine function is continuous.

1.1.4 Affine independence, barycentric coordinates, dimension

A finite set of $(p+1)$ elements of $\mathbb{R}^n$, $\{x^0, x^1, \ldots, x^p\}$, is said to be *affinely independent* if the set $\{(x^1 - x^0), \ldots, (x^p - x^0)\}$ is linearly independent. It should be noticed (and it is easily verified) that this property is independent of the particular choice of x^0 in the set $\{x^0, x^1, \ldots, x^p\}$.

Proposition 1.1.7. *The set $\{x^0, x^1, \ldots, x^p\}$ is affinely independent if and only if $\sum_{k=0}^p \mu_k x^k = 0$ and $\sum_{k=0}^p \mu_k = 0$ imply $\mu_0 = \mu_1 = \ldots = \mu_p = 0$.*

Proof. Suppose that $\{x^0, x^1, \ldots, x^p\}$ is affinely dependent. Then the set $\{(x^1 - x^0), \ldots, (x^p - x^0)\}$ is linearly dependent. Thus there exist p real numbers $\lambda_1, \ldots, \lambda_p$ not all equal to zero such that $\sum_{k=1}^p \lambda_k (x^k - x^0) = 0$. Then $\sum_{k=1}^p \lambda_k x^k + (-\sum_{k=1}^p \lambda_k) x^0 = 0$, which contradicts the property stated in the proposition.

Suppose now that $\{x^0, x^1, \ldots, x^p\}$ is affinely independent and that we have $\sum_{k=0}^p \mu_k x^k = 0$ together with $\sum_{k=0}^p \mu_k = 0$. If some μ_k is not equal to zero, it follows from $\sum_{k=0}^p \mu_k = 0$ that some μ_h, $h \neq k$, is not equal to zero. It follows also from $\sum_{k=0}^p \mu_k = 0$ that $\sum_{k=0}^p \mu_k x^k = 0$ can be rewritten $\sum_{h \neq k} \mu_h (x^h - x^k) = 0$, which contradicts the linear independence of the set $\{(x^h - x^k) \mid h \neq k\}$. □

Proposition 1.1.7 allows us to apply known facts about linear independence to drive results about affine independence in Euclidean spaces. For example, $\mathbb{R}^n$ contains at most $(n+1)$ affinely independent points. If $m < n$, every set of $(m+1)$ affinely independent points of $\mathbb{R}^n$ can be completed in order to obtain a set of $(n+1)$ affinely independent points of $\mathbb{R}^n$. Finally, if the set $\{x^0, x^1, \ldots, x^p\}$ is affinely independent, every point of $\operatorname{aff}(\{x^0, x^1, \ldots, x^p\})$ has a *unique* representation

$$x = \sum_{k=0}^{p} \lambda_k x^k, \ \{\lambda_0, \ldots, \lambda_p\} \subset \mathbb{R}, \sum_{k=0}^{p} \lambda_k = 1. \tag{1.1}$$

We refer to the real coefficients $\lambda_0, \ldots, \lambda_p$ as the *barycentric coordinates* of the point x relative to the affinely independent set $\{x^0, x^1, \ldots, x^p\}$.

On the other hand, in view of Proposition 1.1.7 and using the notations of Proposition 1.1.4, if $V = \{a\} + M$ is an affine space, a set $\{z^1, \ldots, z^p\}$ of p points of M is linearly independent if and only if the set

$$\{a, (a + z^1), \ldots, (a + z^p)\}$$

of $(p+1)$ points of V is affinely independent. The dimension of the unique linear subspace parallel to V is called *dimension* of the affine set V. For a subset A of $\mathbb{R}^n$, the *dimension* of A is by definition the dimension of $\operatorname{aff} A$ (no matter what the dimension of the space in which A is embedded).

Caratheodory's theorem can be precised in the following way:

Proposition 1.1.8. *The convex hull of a p-dimensional set A is precisely the set of all convex combinations of at most $(p+1)$ elements of A.*

Proof. The same as for Proposition 1.1.2. □

1.1.5 Hyperplane

An $(n-1)$-dimensional affine set H of $\mathbb{R}^n$ is called an *affine hyperplane* or *hyperplane.* If H contains the origin, H is an $(n-1)$-dimensional linear subspace of $\mathbb{R}^n$. If $a \in H$, $H = \{a\} + M$, where M is the $(n-1)$-dimensional linear subspace of $\mathbb{R}^n$ parallel to H.

Hyperplanes, and more generally affine sets, can be viewed as sets of solutions to linear equations.

Proposition 1.1.9. *Let H be a subset of $\mathbb{R}^n$. The set H is a hyperplane if and only if there exists $p \in \mathbb{R}^n \setminus \{0\}$ and $\lambda \in \mathbb{R}$ such that $H = \{x \in \mathbb{R}^n \mid p \cdot x = \lambda\}$. Moreover, p and λ are unique up to a same scalar factor.*

Proof. Set $H = \{a\} + M$, where M is the $(n-1)$-dimensional linear subspace of $\mathbb{R}^n$ parallel to H. Recall that $\mathbb{R}^n = M \oplus M^\perp$ where $M^\perp$ is the linear subspace orthogonal to M, that is $M^\perp = \{q \in \mathbb{R}^n \mid q \cdot x = 0, \ \forall x \in M\}$. Clearly, $M^\perp$ is one-dimensional and has a unique, up to a scalar factor, generator p. Moreover, $M = (M^\perp)^\perp$. Thus,

$$z \in H \Leftrightarrow (z-a) \in M \Leftrightarrow p \cdot (z-a) = 0 \Leftrightarrow p \cdot x = p \cdot a$$

so that $H = \{x \in \mathbb{R}^n \mid p \cdot x = \lambda\}$ with λ defined by $\lambda = p \cdot a$.

Conversely, if $H = \{x \in \mathbb{R}^n \mid p \cdot x = \lambda\}$ and if $a \in H$, then $p \cdot a = \lambda$ and $H = \{a\} + \{x \in \mathbb{R}^n \mid p \cdot x = 0\}$. Thus H is a hyperplane. □

The nonzero vector p in the previous proposition is said to be a *normal vector* of H. If H is parallel to the $(n-1)$-dimensional subspace M, the normal vector is the same for every hyperplane parallel to M.

Corollary 1.1.1. *Let H be a subset of $\mathbb{R}^n$. The set H is a hyperplane if and only if there exists a linear functional f and $\lambda \in \mathbb{R}$ such that $H = \{x \in \mathbb{R}^n \mid f(x) = \lambda\}$. f and λ are unique up to a same scalar factor.*

The proof of the following proposition is left for the reader.

Proposition 1.1.10. *Let $b \in \mathbb{R}^m$ and A be a $(m \times n)$-real matrix. the set*

$$V = \{x \in \mathbb{R}^n \mid Ax = b\}$$

is an affine set of $\mathbb{R}^n$. If $V \neq \emptyset$ and $\operatorname{rank} A = p$, *then* $\dim V = (n-p)$. *Conversely, every nonempty affine set of $\mathbb{R}^n$ can be represented in this way.*

Corollary 1.1.2. *Every affine set of $\mathbb{R}^n$ is the intersection of a finite number of hyperplanes.*

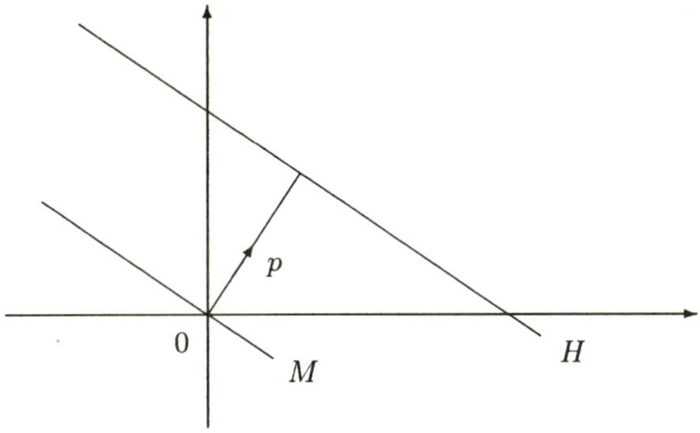

Fig. 1.1.

1.1.6 Convex cone and convex conic hull

A subset K of $\mathbb{R}^n$ is a *cone* with *vertex* 0 (or simply a *cone*) if K is closed under positive scalar multiplication, that is, if $\lambda x \in K$ when $x \in K$ and $\lambda > 0$. Such a set is a union of half-lines emanating from the origin. The origin may or may not be included. In the first case, the cone K is said to be *pointed.* A *convex cone* is a cone which is a convex set.

Proposition 1.1.11. *A subset K of $\mathbb{R}^n$ is a convex cone if and only if $x \in K$, $y \in K$ and $\lambda > 0$ imply $x + y \in K$ and $\lambda x \in K$.*

Proof. Elementary. □

Linear subspaces of $\mathbb{R}^n$ are convex cones. Any intersection of convex cones is a convex cone. With any arbitrary subset A of $\mathbb{R}^n$, we can associate a convex cone, called the *convex cone generated by* A (*convex conic hull* of A) and denoted by $\mathrm{K}(A)$, which is the intersection of all convex cones of $\mathbb{R}^n$ containing A (the smallest convex cone of $\mathbb{R}^n$ containing A). For every finite set $\{x^1, \ldots, x^p\}$ of p elements of A and for every set of p positive (resp. nonnegative) real coefficients $\lambda_k > 0$ (resp. $\lambda_k \geq 0$), $k = 1, \ldots, p$, let us call $\sum_{k=1}^p \lambda_k x^k$ *positive* (resp. *nonnegative*) *linear combination* of $\{x^1, \ldots, x^p\}$.

Proposition 1.1.12. *The convex cone generated by a subset A of $\mathbb{R}^n$ is precisely the set of all positive linear combinations of finitely many elements of A.*

Proof. Left for the reader (see the exercise 1.14). □

Equivalently, the pointed convex cone generated by A is the set of all nonnegative linear combinations of finitely many members of A.

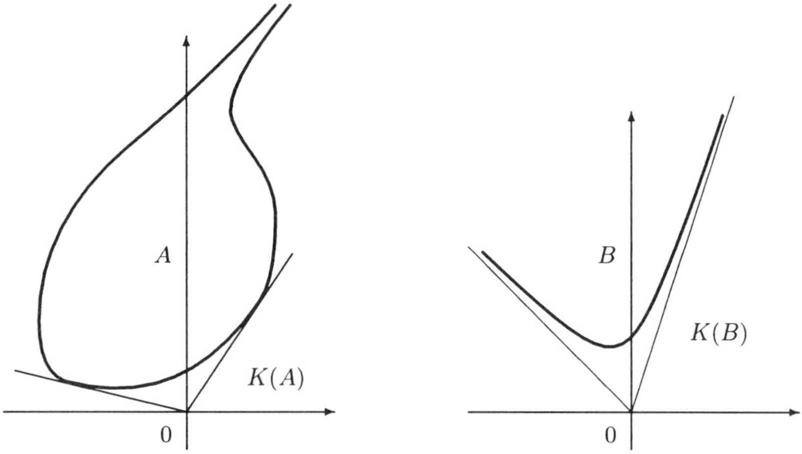

Fig. 1.2. Convex cone generated by a set

1.2 Topological properties of convex sets

In what follows, the topology of $\mathbb{R}^n$ is the topology associated to the Euclidean metric. As usual, the topology on every subset of $\mathbb{R}^n$ is the topology induced on the subset by the topology of $\mathbb{R}^n$.

1.2.1 Closure and interior of a convex set

Let us first remark that every p-dimensional linear subspace of $\mathbb{R}^n$, being the kernel of $(n-p)$ linearly independent (continuous) linear functionals, is a closed subset of $\mathbb{R}^n$. Likewise, every affine set, being the translate of a linear subspace, is a closed subset of $\mathbb{R}^n$.

Proposition 1.2.1. *The closure of a convex subset of $\mathbb{R}^n$ is a convex subset of the same dimension.*

Proof. Let $A \subset \mathbb{R}^n$ be convex. We first show that $\overline{A}$ is convex: let x and y be in $\overline{A}$, λ be a real number satisfying: $0 < \lambda < 1$ and $\varepsilon > 0$. Since x and y belong to $\overline{A}$, there exist x_ε and y_ε in A such that $\|x_\varepsilon - x\| < \varepsilon$ and $\|y_\varepsilon - y\| < \varepsilon$. Then $\|(\lambda x_\varepsilon + (1-\lambda)y_\varepsilon) - (\lambda x + (1-\lambda)y)\| \leq \lambda\|x_\varepsilon - x\| + (1-\lambda)\|y_\varepsilon - y\| < \varepsilon$. Since A is convex, $\lambda x_\varepsilon + (1-\lambda)y_\varepsilon \in A$ and thus $\lambda x + (1-\lambda)y \in \overline{A}$.

Let V be the affine hull of A. Since $A \subset V$ and V is closed, $\overline{A} \subset V$. It is easily seen that V is also the affine hull of $\overline{A}$. The sets A and $\overline{A}$ having same affine hull have same dimension. □

Defining the *closed convex hull* of a subset A of $\mathbb{R}^n$, $\overline{\mathrm{co}}\, A$, as the smallest closed convex set that contains A, that is the intersection of all closed convex subsets of $\mathbb{R}^n$ containing A, one gets the following corollary:

Corollary 1.2.1. *The closed convex hull of a subset A of $\mathbb{R}^n$ is the closure of the convex hull of A:* $\overline{\mathrm{co}}\,A = \overline{\mathrm{co}\,A}$.

Proof. In view of Proposition 1.2.1, $\overline{\mathrm{co}\,A}$ is convex, closed and contains A, thus contains $\overline{\mathrm{co}}\,A$. Conversely, the closed convex set $\overline{\mathrm{co}}\,A$ contains $\mathrm{co}\,A$, thus contains $\overline{\mathrm{co}\,A}$. □

Remark 1.2.1. It should be emphasized that the equality $\mathrm{co}\,\overline{A} = \overline{\mathrm{co}\,A}$ does not hold in general (see below in Subsection 1.2.3). Corollary 1.2.1 does not mean that the order of the operations of taking closure and convex hull can be reversed.

Proposition 1.2.2. *If a convex set A has a nonempty interior* $\mathrm{i}(A) \neq \emptyset$, *then for every α: $0 \leq \alpha < 1$ one has:* $\alpha\overline{A} + (1-\alpha)\,\mathrm{i}(A) \subset \mathrm{i}(A)$.

Proof. Let α be such that $0 \leq \alpha < 1$. Since $\alpha\overline{A} + (1-\alpha)\,\mathrm{i}(A)$ is an open set (see the exercise 1.5), it suffices to show that $\alpha\overline{A} + (1-\alpha)\,\mathrm{i}(A) \subset A$. If $z \in \mathrm{i}(A)$, then $(1-\alpha)\big(\mathrm{i}(A) - \{z\}\big)$ is an open set containing 0. One deduces (see the same exercise):

$$\alpha\overline{A} = \overline{\alpha A} \subset \alpha A + (1-\alpha)\big(\mathrm{i}(A) - \{z\}\big)$$

thus $\alpha\overline{A} + (1-\alpha)\{z\} \subset \alpha A + (1-\alpha)\,\mathrm{i}(A)$. The previous inclusion being true for every z of $\mathrm{i}(A)$, one has: $\alpha\overline{A} + (1-\alpha)\,\mathrm{i}(A) \subset A$. □

Remark 1.2.2. The previous proposition is most often stated as follows: Given $z \in \mathrm{i}(A)$ and $x \in \overline{A}$, every point of the line segment $[z, x)$ belongs to $\mathrm{i}(A)$.

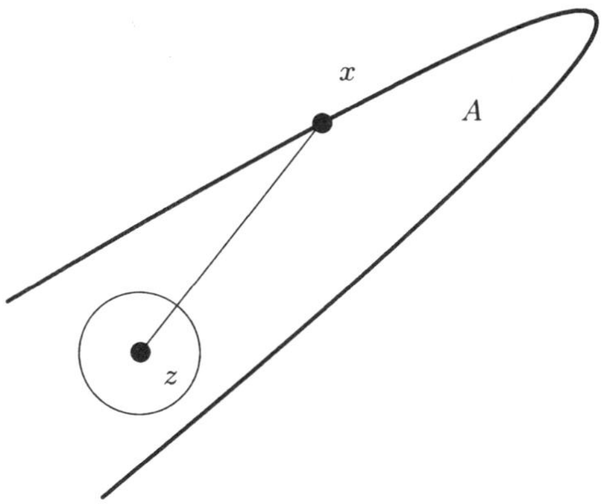

Fig. 1.3.

Corollary 1.2.2. *The interior of a convex set A is a convex set. If* $\mathrm{i}(A) \neq \emptyset$, *then* $\mathrm{i}(A) = \mathrm{i}(\overline{A})$ *and* $\overline{A} = \overline{\mathrm{i}(A)}$.

Proof. The first assertion follows immediately from Proposition 1.2.2. The inclusions $\mathrm{i}(A) \subset \mathrm{i}(\overline{A})$ and $\overline{\mathrm{i}(A)} \subset \overline{A}$ are also immediate.

Let $x \in \mathrm{i}(\overline{A})$ and $z \in \mathrm{i}(A)$. Since $x \in \mathrm{i}(\overline{A})$, there exists $u \in \overline{A}$ verifying $u = x - \lambda(z - x)$ for some $\lambda > 0$. Then $x = (1/(1+\lambda))u + (\lambda/(1+\lambda))z$ belongs to the open line segment (z, u) and it follows from Proposition 1.2.2 that $x \in \mathrm{i}(A)$ (see Fig. 1.4).

On the other hand, let $u \in \overline{A}$ and and $z \in \mathrm{i}(A)$. All the points of the line segment $[z, u)$ belong to $\mathrm{i}(A)$, hence $u \in \overline{\mathrm{i}(A)}$ thus $\overline{A} \subset \overline{\mathrm{i}(A)}$. □

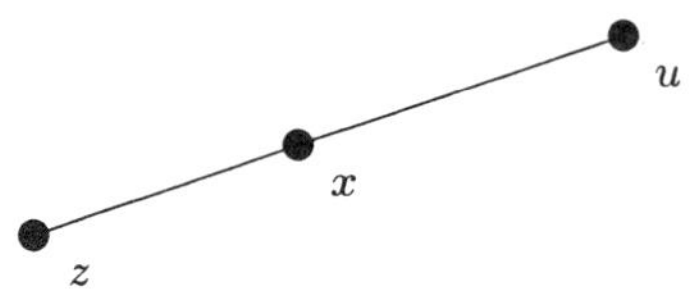

$x \in \mathrm{i}(\overline{A})$; on the other hand $u \in \overline{A}$ and $z \in \mathrm{i}(A)$ imply that $x \in \mathrm{i}(A)$

Fig. 1.4. $\mathrm{i}(\overline{A}) \subset \mathrm{i}(A)$

Corollary 1.2.3. *The convex hull of an open subset of* $\mathbb{R}^n$ *is an open subset of* $\mathbb{R}^n$.

Proof. For an open set A, let $\operatorname{co} A$ be its convex hull. Since A is open, $A \subset \mathrm{i}(\operatorname{co} A)$. By the previous corollary, $\mathrm{i}(\operatorname{co} A)$ is convex and then $\operatorname{co} A \subset \mathrm{i}(\operatorname{co} A)$. Since one obviously has $\mathrm{i}(\operatorname{co} A) \subset \operatorname{co} A$, it follows that $\mathrm{i}(\operatorname{co} A) = \operatorname{co} A$, hence that $\operatorname{co} A$ is open. □

Let us now prove two lemmas.

Lemma 1.2.1. *Let* $\{x^0, x^1, \dots, x^n\}$ *be a set of* $n+1$ *affinely independent points of* $\mathbb{R}^n$. *The set*

$$\{z \in \mathbb{R}^n \mid z = \sum_{k=0}^{n} \alpha_k x^k, \alpha_k > 0\ \forall k = 0, \dots, n \text{ and } \sum_{k=0}^{n} \alpha_k = 1\}$$

is an open subset of $\mathbb{R}^n$.

Proof. For every x of $\mathbb{R}^n$, let $\alpha_k(x)$, $k = 0, \dots n$ denote the barycentric coordinates of x relative to the affinely independent set $\{x^0, x^1, \dots, x^n\}$. Using (1.1), one easily verifies that the maps $\alpha_k : \mathbb{R}^n \to \mathbb{R}$ are affine (hence continuous) maps. Moreover, the set

$$\{z \in \mathbb{R}^n \mid z = \sum_{k=0}^{n} \alpha_k x^k, \alpha_k > 0\ \forall k = 0, \dots, n \text{ and } \sum_{k=0}^{n} \alpha_k = 1\}$$

coincides with the intersection $\bigcap_{k=0}^{n} (\alpha_k)^{-1}(]0,1[)$ and this set is open as a finite intersection of open subsets of $\mathbb{R}^n$. □

We will call *p-simplex* the convex hull of $(p+1)$ affinely independent points of $\mathbb{R}^n$: $x^0, x^1, \ldots, x^p$. The points $x^0, x^1, \ldots, x^p$ are called *vertices* of the p-simplex. The previous lemma shows that a n-simplex has a nonempty interior.

Lemma 1.2.2. *A convex subset of $\mathbb{R}^n$ is p-dimensional if and only if it contains a set of $(p+1)$ affinely independent points of $\mathbb{R}^n$ and no set of $(k+1)$ affinely independent points of $\mathbb{R}^n$ with $k > p$.*

Proof. Let A be a convex subset of $\mathbb{R}^n$. Let us first remark that A contains sets of affinely independent points and that no such set has more than $(n+1)$ elements. Let us define the maximal number of such a set and assume that $(p+1)$ is this number and that $\{x^0, \ldots, x^p\}$ is such a set. Let $\operatorname{aff} A$ and $\operatorname{aff}(\{x^0, \ldots, x^p\})$ denote the affine hulls of A and $\{x^0, \ldots, x^p\}$ respectively. We claim that $\operatorname{aff} A = \operatorname{aff}(\{x^0, \ldots, x^p\})$.

To prove the claim, we have just to show that $\operatorname{aff} A \subset \operatorname{aff}(\{x^0, \ldots, x^p\})$. Indeed, if $x \in A$, the set $\{x, x^0, \ldots, x^p\}$ is affinely dependent; therefore there exist λ and λ_k, $k = 0, \ldots, p$ not all equal to 0 and with $\lambda + \sum_{k=0}^p \lambda_k = 0$ such that $\lambda x + \sum_{k=0}^p \lambda_k x^k = 0$. Since the set $\{x^0, \ldots, x^p\}$ is affinely independent, $\lambda \neq 0$ and $\lambda_k \neq 0$ for some k. So x can be written $x = \sum_{k=0}^p \mu_k x^k$ with $\sum_{k=0}^p \mu_k = 1$. In other words, $x \in \operatorname{aff} A$, which proves that $A \subset \operatorname{aff}(\{x^0, \ldots, x^p\})$, thus $\operatorname{aff} A \subset \operatorname{aff}(\{x^0, \ldots, x^p\})$.

We now have that $\operatorname{aff} A = \operatorname{aff}(\{x^0, \ldots, x^p\})$. The dimension of A is the dimension of $\operatorname{aff}(\{x^0, \ldots, x^p\})$ and is thus equal to p. □

We are now ready to state the main result of this subsection.

Proposition 1.2.3. *In $\mathbb{R}^n$, a convex set A has a nonempty interior if and only if it is n-dimensional.*

Proof. If A is n-dimensional, A contains a set of $n+1$ affinely independent points $\{x^0, x^1, \ldots, x^n\}$ (Lemma 1.2.2) and the set $\{x \in \mathbb{R}^n \mid x = \sum_{k=0}^n \alpha_k x^k, \alpha_k > 0, k = 0, \ldots, n, \text{ and } \sum_{k=0}^n \alpha_k = 1\}$ is nonempty, open (Lemma 1.2.1) and contained in A, hence in $\mathrm{i}(A)$.

If A is p-dimensional, with $p < n$, then $A \subset \operatorname{aff} A$ and $\dim \operatorname{aff} A = p$. Let H be an hyperplane containing $\operatorname{aff} A$. By Proposition 1.1.9, $H = \{x \in \mathbb{R}^n \mid p \cdot x = \lambda\}$ for some $p \in \mathbb{R}^n$ and $\lambda \in \mathbb{R}$. If $x \in H$ and $z \notin H$ (for example if $p \cdot z < p \cdot x = \lambda$), it is easily seen that the whole line segment $(x, z]$ is contained in $\mathbb{R}^n \setminus H$. Hence no ball with center x can be contained in H and x does not belong to the interior of A. We have thus proved that $x \in H$ implies $x \notin \mathrm{i}(A)$. A fortiori, $x \in A$ implies $x \notin \mathrm{i}(A)$, hence $\mathrm{i}(A) = \emptyset$. □

1.2.2 Relative interior and relative boundary of a convex set

The concepts introduced in this subsection are motivated by the fact that if $p < n$, p-dimensional convex subsets of $\mathbb{R}^n$ have an empty interior. The *relative interior* of a convex set A in $\mathbb{R}^n$, which we denote by $\operatorname{ri}(A)$, is the interior of A in aff A. Thus ri(A) is the set of the points x of A such that there exists a ball with center x and radius ε whose intersection with aff A is contained in A. If A is n-dimensional, aff $A = \mathbb{R}^n$ and the relative interior of A coincides with its interior. A is said to be *relatively open* if ri$(A) = A$; in particular, an affine set is relatively open. In view of the closedness of aff A, notice here that a subset of aff A is closed in aff A if and only if it is closed and that the closure in aff A of a subset of aff A coincides with its closure. The set difference $\overline{A} \setminus \text{ri}(A)$ is called the *relative boundary* of the convex set A and denoted Bd$^\text{r}(A)$. Once again, if the convex set A is n-dimensional, the relative boundary of A coincides with its boundary $\partial A = \overline{A} \setminus \text{i}(A)$.

The following lemma is an analogue of Lemma 1.2.1.

Lemma 1.2.3. *Let $\{x^0, x^1, \ldots, x^p\}$ be a set of $(p+1)$ affinely independent points of $\mathbb{R}^n$. The set*

$$\{x \in \mathbb{R}^n \mid x = \sum_{k=0}^{p} \alpha_k x^k, \alpha_k > 0 \ \forall k = 0, \ldots, p \text{ and } \sum_{k=0}^{p} \alpha_k = 1\}$$

is relatively open.

Proof. It is easily seen that aff$(\{x^0, x^1, \ldots, x^p\})$ is the affine hull of this set. As in the proof of Lemma 1.2.1, if $\alpha_k(x)$, $k = 0, \ldots, p$, denote the barycentric coordinates of a point $x \in \text{aff}(\{x^0, x^1, \ldots, x^p\})$, the maps α_k : aff$(\{x^0, x^1, \ldots, x^p\}) \to \mathbb{R}$ are affine (hence continuous) functions. The intersection $\bigcap_{k=0}^{p} (\alpha_k)^{-1}(]0, 1[)$ is open in aff$(\{x^0, x^1, \ldots, x^p\})$, thus relatively open. □

We are now ready to prove:

Proposition 1.2.4. *In $\mathbb{R}^n$, every nonempty convex set has a nonempty relative interior.*

Proof. Let A be a p-dimensional convex set. By Lemma 1.2.2, A contains a set $\{x^0, x^1, \ldots, x^p\}$ of $(p+1)$ affinely independent points and aff A = aff$(\{x^0, x^1, \ldots, x^p\})$. By Proposition 1.2.3, the nonempty set $\{x \in \mathbb{R}^n \mid x = \sum_{k=0}^{p} \alpha_k x^k, \alpha_k > 0 \ \forall k = 0, \ldots, p$ and $\sum_{k=0}^{p} \alpha_k = 1\}$ is relatively open and contained in the relative interior of A which is thus nonempty. □

The following properties of closures and relative interiors of convex sets extend Proposition 1.2.2 and its corollaries.

Proposition 1.2.5. *Let A be convex. Then for every α : $0 \le \alpha < 1$, one has: $\alpha\overline{A} + (1-\alpha)\,\text{ri}(A) \subset \text{ri}(A)$.*

Proof. Let us first prove the weaker assertion: $\alpha A + (1-\alpha)\,\text{ri}(A) \subset \text{ri}(A)$. Indeed, let $z \in \text{ri}(A)$, $y \in A$ and $x = \alpha y + (1-\alpha)z$. One has: $x - y = (1-\alpha)(z-y)$. Since $z \in \text{ri}(A)$, there exists r such that $\big(B_0(z,r) \cap \text{aff } A\big) \subset A$,

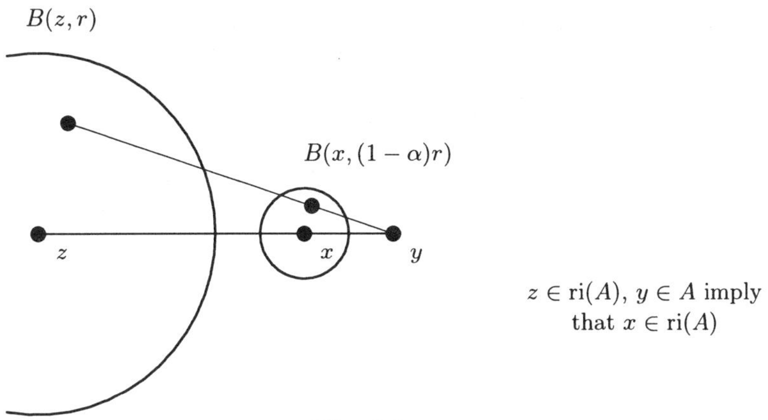

Fig. 1.5.

where $B_0(z, r)$ is the open ball with center z and radius r. It is easily verified that $\big(B_0(x,(1-\alpha)r) \cap \operatorname{aff} A\big) \subset A$, which proves that $x \in \operatorname{ri}(A)$.

We then mimic the arguments of the proof of Proposition 1.2.2. One can see $\operatorname{ri}(A)$ as the intersection with aff A of an open set U of $\mathbb{R}^n$. If $z \in \operatorname{ri}(A)$, $U - \{z\}$ is an open set containing 0. In view of the inclusion $\alpha\overline{A} = \overline{\alpha A} \subset \alpha A + (1-\alpha)(U - \{z\})$, we get $\alpha\overline{A} + (1-\alpha)\{z\} \subset \alpha A + (1-\alpha)U$. Observing that $\alpha\overline{A} + (1-\alpha)\{z\} \subset \operatorname{aff} A$, we then have: $\alpha\overline{A} + (1-\alpha)\{z\} \subset \big(\alpha A + (1-\alpha)U\big) \cap \operatorname{aff} A \subset \alpha A + (1-\alpha)(U \cap \operatorname{aff} A) = \alpha A + (1-\alpha)\operatorname{ri}(A)$. By the first part of the proof, $\alpha A + (1-\alpha)\operatorname{ri}(A) \subset \operatorname{ri}(A)$, which ends the proof. □

Corollary 1.2.4. *The relative interior of a convex set is a convex set of same dimension. Moreover,* $\operatorname{ri}(\overline{A}) = \operatorname{ri}(A)$ *and* $\overline{\operatorname{ri}(A)} = \overline{A}$.

Proof. By Proposition 1.2.5, $\operatorname{ri}(A)$ is a convex set. If A is p-dimensional and if $\{x^0, x^1, \ldots, x^p\}$ is a set of $(p+1)$ affinely independent points of A (Lemma 1.2.2), the obviously p-dimensional set $\{x \in \mathbb{R}^n \mid x = \sum_{k=0}^p \alpha_k x^k,\ \alpha_k > 0,\ \forall k = 0, \ldots, p \text{ and } \sum_{k=0}^p \alpha_k = 1\}$ is contained in $\operatorname{ri} A$. The rest of the proof is the same as for Corollary 1.2.2. □

Corollary 1.2.5. *Let A be a nonempty convex set. In order that $x \in \operatorname{ri}(A)$, it is necessary and sufficient that every line passing through x and any point y of A contain an open line segment contained in A and containing x.*

Proof. The condition is obviously necessary. Let us show that it is sufficient. Let x verify the condition and let $y \in \operatorname{ri}(A)$ which is nonempty (Proposition 1.2.4). The line passing through x and y contains an open line segment containing x and contained in A. Let z in this line segment at the opposite side of y relative to x. It then follows from Proposition 1.2.5 that $x \in \operatorname{ri}(A)$. □

Corollary 1.2.6. *Let A be a convex set with a nonempty interior. In order that $x \in \operatorname{i}(A)$, it is necessary and sufficient that every line passing through*

x and any point y of A contain an open line segment contained in A and containing x.

Proof. Corollary 1.2.6 is a particular case of Corollary 1.2.5. □

It is worth noticing that if $x \in \mathrm{ri}(A)$, then every line passing through x and contained in aff A contains an open line segment contained in A and containing x. Corollary 1.2.5 can then easily be rephrased as:

Corollary 1.2.7. *Let A be a nonempty convex set. Let F be the linear subspace parallel to* aff A. *In order that $x \in \mathrm{ri}(A)$, it is necessary and sufficient that for every $u \in F$, there exists $\lambda > 0$ such that $x + \lambda u \in A$.*

Proof. It suffices to note that if $x \in A$, then $u \in F$ if and only if $x + u \in \mathrm{aff}\, A$. Corollary 1.2.7 follows from Corollary 1.2.5 and from the convexity of A. □

1.2.3 Convex hull of a compact set

The convex hull of a closed set A need not be closed. Consider, for example, the set $A = \{x \in \mathbb{R}^2 \mid |x_1|x_2 \geq 1\}$ and its convex hull $\mathrm{co}\, A = \{x \in \mathbb{R}^2 \mid x_2 > 0\}$ (see Fig. 1.6). In the exercise 1.22, an example, borrowed from Fenchel (1953), shows that for a closed cone, the equality $\mathrm{co}\, C = \overline{\mathrm{co}\, C}$ does not hold in general.

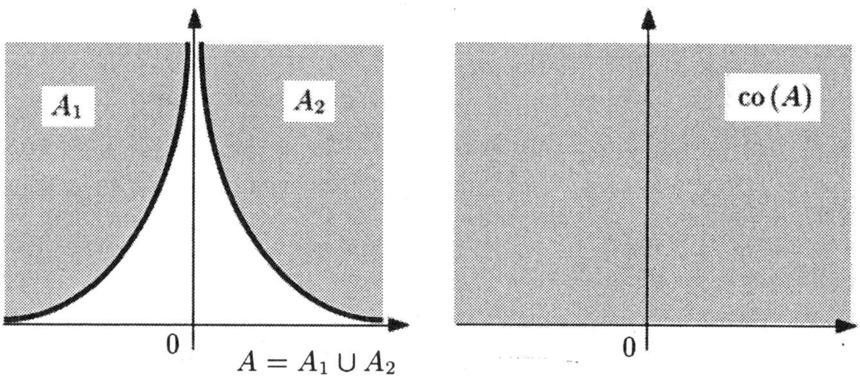

Fig. 1.6. A is closed and co A is open

Things are different if the closed set is bounded.

Proposition 1.2.6. *The convex hull of finitely many points is a compact subset of $\mathbb{R}^n$.*

More generally,

Proposition 1.2.7. *In $\mathbb{R}^n$, the convex hull of a compact set is a compact set.*

Proof. Let $\Delta_n = \{\lambda \in \mathbb{R}^{n+1} \mid 0 \leq \lambda_k,\ k = 0, \ldots, n \text{ and } \sum_{k=0}^{n} \lambda_k = 1\}$ be the unit-simplex of $\mathbb{R}^{n+1}$. Δ_n is a closed and bounded, hence compact, subset of $\mathbb{R}^{n+1}$. The Caratheodory theorem (Proposition 1.1.2) shows that $\operatorname{co} A$ is the image of $\Delta_n \times A \times \ldots \times A$ by the map

$$(\lambda, x^0, \ldots, x^n) \to \sum_{k=0}^{n} \lambda_k x^k.$$

This map is obviously continuous and $\Delta_n \times A \times \ldots \times A$ is a compact set in $\mathbb{R}^{n+1} \times \mathbb{R}^n \times \ldots \times \mathbb{R}^n$ endowed with the product topology. Thus the set $\operatorname{co} A$ is a compact subset of $\mathbb{R}^n$. □

Exercises

The five first exercises recall some useful properties of the topology of $\mathbb{R}^n$. The other ones deal with the material of Chapter 1.

Exercise 1.1. Prove the continuity of the following maps:
a) $x \to \|x\|$ (map defined on $\mathbb{R}^n$ with values in $\mathbb{R}_+$).
b) If a is a point of $\mathbb{R}^n$, $x \to \mathrm{d}(a, x) = \|x - a\|$.
c) If a is a point of $\mathbb{R}^n$, $x \to a + x$ (map defined on $\mathbb{R}^n$ with values in $\mathbb{R}^n$).
d) If $\lambda \neq 0$, $x \to \lambda x$ (map defined on $\mathbb{R}^n$ with values in $\mathbb{R}^n$).
e) $x = (x_1, \ldots, x_n) \to x_i$ (i^{th} projection from $\mathbb{R}^n$ to $\mathbb{R}$).
f) Every linear map defined on $\mathbb{R}^n$ with values in $\mathbb{R}^p$.

Exercise 1.2. On $\mathbb{R}^n$, besides the Euclidean norm, one can consider the following norms: $\|x\|_\infty = \sup_i |x_i|$, $\|x\|_1 = \sum_{i=1}^{n} |x_i|$.
Verify the following relations:

$$\|x\|_\infty \leq \|x\| \leq \|x\|_1 \leq n\|x\|_\infty.$$

Prove that these norms define the same topology on $\mathbb{R}^n$.

Exercise 1.3. Prove the compactness of the following subsets of $\mathbb{R}^n$:
a) $\Delta_{n-1} = \{x \in \mathbb{R}^n_+ \mid \sum_{i=1}^{n} x_i = 1\}$ (unit-simplex of $\mathbb{R}^n$),
b) $B_n = \overline{B(0,1)} = \{x \in \mathbb{R}^n \mid \|x\| \leq 1\}$ (unit-ball of $\mathbb{R}^n$),
c) $S_{n-1} = \overline{B(0,1)} \setminus B(0,1) = \{x \in \mathbb{R}^n \mid \|x\| = 1\}$ (unit-sphere of $\mathbb{R}^n$),
d) $P_{ab} = \{x \in \mathbb{R}^n \mid a_i \leq x_i \leq b_i,\ \forall i = 1, \ldots, n\}$, $a \in \mathbb{R}^n$, $b \in \mathbb{R}^n$.

Exercise 1.4. For $x \in \mathbb{R}^n$, $A \subset \mathbb{R}^n$, $B \subset \mathbb{R}^n$, one defines:
$d(x, A) = \inf\{d(x, y) \mid y \in A\}$; $d(A, B) = \inf\{d(x, y) \mid x \in A,\ y \in B\}$.
Prove that the map $x \to d(x, A)$ is a continuous function $\mathbb{R}^n \to \mathbb{R}$.

Recalling that $\overline{A}$ denotes the closure of A, prove that $d(x, A) = 0$ if and only if $x \in \overline{A}$.
If A is compact, B is closed and $A \cap B = \emptyset$, prove that $d(A, B) > 0$.
Under the same assumptions, prove that there exist two open sets U and V such that: $A \subset U$, $B \subset V$, $U \cap V = \emptyset$.
Finally, if K is compact, U is open and $K \subset U$, show that the exists a real number $\varepsilon > 0$ such that $K + \varepsilon\overline{B(0,1)} \subset U$

Exercise 1.5. Prove the following assertions for $\alpha \in \mathbb{R}$, A, B, C, U, V in $\mathbb{R}^n$:
a) $\alpha \neq 0$ and C open imply αC open.
b) $\alpha \neq 0$ and C closed imply αC closed.
c) $\alpha\overline{C} = \overline{\alpha C}$.
d) If U is open then whatever be V, $U + V$ is open.
e) If V is an open set containing the origin, then $\overline{C} \subset C + V$.
f) If A and B are compact, then $(A + B)$ is compact.
g) If A is compact and B is closed, then $(A + B)$ is closed.
h) $\overline{A} + \overline{B} \subset \overline{A + B}$; the converse inclusion is true if either A or B is bounded. Give an example where $\overline{A + B} \not\subset \overline{A} + \overline{B}$.
(Recall that $U + V := \{z \in \mathbb{R}^n \mid z = x + y, x \in U, y \in V\}$; $\alpha U := \{z \in \mathbb{R}^n \mid z = \alpha x, x \in U\}$)

Exercise 1.6. Show that the following sets are convex:
every interval of $\mathbb{R}$,
$\{(x, y) \in \mathbb{R}^2 \mid x^2 + y^2 < 4\}$,
$\{(x, y) \in \mathbb{R}^2 \mid |x| + |y| \leq 2\}$,
$\{(x, y) \in \mathbb{R}^2 \mid \max\{|x|, |y|\} \leq 2\}$,
the set of the $(n \times n)$-matrices with elements ≥ 0.

Exercise 1.7. Show that every interval of $\mathbb{R}$ is a line segment of $\mathbb{R}$ and that the only nonempty convex subsets of $\mathbb{R}$ are the intervals.

Exercise 1.8. Let A be a subset of $\mathbb{R}^n$. If A is closed, prove that A is convex if and only if $(1/2)(x + y) \in A$ whenever $x \in A$ and $y \in A$. What if A is not assumed to be closed?

Exercise 1.9. Let U and V be two subsets of $\mathbb{R}^n$.
If U and V are convex, show that $U + V$ is convex and that αU is convex (whatever be $\alpha \in \mathbb{R}$).
Show that: $\operatorname{co} U + \operatorname{co} V \subset \operatorname{co}(\operatorname{co} U + V)$; $\operatorname{co} U + V \subset \operatorname{co}(U + V)$.
Prove then that $\operatorname{co}(U + V) = \operatorname{co} U + \operatorname{co} V$.
If $\overline{\operatorname{co}}\, U$ is compact, show that $\overline{\operatorname{co}}(U + V) = \overline{\operatorname{co}}\, U + \overline{\operatorname{co}}\, V$.

Exercise 1.10. Prove that for convex subsets A and B of $\mathbb{R}^n$,
$\operatorname{co}(A \cup B) = \{z = \lambda x + (1 - \lambda)y \mid x \in A,\ y \in B,\ 0 \leq \lambda \leq 1\}$.
More generally, prove that for a finite collection $(A^i)_{i=1}^p$ of convex subsets of $\mathbb{R}^n$,
$\operatorname{co}(\bigcup_{i=1}^p A^i) = \{z = \sum_{i=1}^p \lambda_i x^i \mid \lambda_i \geq 0,\ x^i \in A^i, i = 1, \ldots p,\ \sum_{i=1}^p \lambda_i = 1\}$.

Exercise 1.11. Let $U^i, i = 1, 2, \ldots$ be convex sets.
Show that if for every j, $U^j \subset U^{j+1}$, then $\bigcup_{j=1}^{\infty} U^j$ is convex.
Show that $\bigcup_{k=1}^{\infty} \bigcap_{j=k}^{\infty} U^j$ is a convex set.

Exercise 1.12. Let $\mathcal{C}$ be the set of all convex subsets of $\mathbb{R}^n$, partially ordered by inclusion. Show that for every indexed family $(C^i)_{i \in I}$ of elements of $\mathcal{C}$, one can define (in $\mathcal{C}$) $\inf_{i \in I} C^i$ and $\sup_{i \in I} C^i$ (i.e., that $\mathcal{C}$ is a complete lattice).

Exercise 1.13. In $\mathbb{R}^2$, consider the triangle with vertices x^0, x^1, x^2. Let y be the midpoint of the side opposite x^0 and let z be the intersection of the three medians. What are (relative to the set of affinely independent points $\{x^0, x^1, x^2\}$) the barycentric coordinates of respectively: x^0, x^1, x^2, y, z?

Exercise 1.14. Prove the proposition 1.1.12.
If A is a convex subset of $\mathbb{R}^n$ which does not contain 0, show that the convex cone (with vertex 0) generated by A is precisely the set $C = \bigcup_{\lambda > 0} \lambda A$.

Exercise 1.15. Let C be a nonempty convex cone (with vertex 0) in $\mathbb{R}^n$.
a) Show that the set $C - C$ is the linear subspace generated by C. If $0 \in C$, show that $L := C \bigcap (-C)$ is the biggest linear subspace contained in C.
b) In this case, denote by $L^\perp$ the linear subspace of $\mathbb{R}^n$ orthogonal to L. Recall that $L \cap L^\perp = \{0\}$ and that every x of $\mathbb{R}^n$ has a unique decomposition $x = y + z$, $y \in L$, $z \in L^\perp$. Show that $K := C \cap L^\perp$ is a convex cone verifying $K \cap (-K) = \{0\}$ and that $C = L + K$.

Exercise 1.16 (Gauge of a convex set). Let A be a convex subset of $\mathbb{R}^n$ such that $0 \in \mathrm{i}(A)$. One defines the function $\gamma_A : \mathbb{R}^n \to \mathbb{R}_+$

$$\gamma_A(x) = \inf\{\lambda > 0 \mid x \in \lambda A\}$$

and the sets

$$\begin{aligned} W(\gamma_A, 1) &= \{x \in \mathbb{R}^n \mid \gamma_A(x) < 1\}, \\ V(\gamma_A, 1) &= \{x \in \mathbb{R}^n \mid \gamma_A(x) \le 1\}. \end{aligned}$$

a) Show that γ_A is well-defined and that for every $\alpha \ge 0$, and x, y in $\mathbb{R}^n$:

$$\begin{aligned} \gamma_A(\alpha x) &= \alpha \gamma_A(x), \\ \gamma_A(x + y) &\le \gamma_A(x) + \gamma_A(y). \end{aligned}$$

b) Show that $W(\gamma_A, 1) \subset A \subset V(\gamma_A, 1)$ and that

$$\begin{aligned} \gamma_A(x) < 1 &\iff x \in \mathrm{i}(A), \\ \gamma_A(x) \le 1 &\iff x \in \overline{A}, \\ \gamma_A(x) = 1 &\iff x \in \partial A := \overline{A} \setminus \mathrm{i}(A). \end{aligned}$$

Deduce from the previous results that

$$A = W(\gamma_A, 1) \iff A \text{ is open},$$
$$A = V(\gamma_A, 1) \iff A \text{ is closed}.$$

If moreover A is compact, then $\forall x \neq 0$, $\gamma_A(x) > 0$ and $(1/\gamma_A(x))x \in A \setminus \mathrm{i}(A)$.

Exercise 1.17 (Recession cone of a convex set). Given C, a nonempty convex subset of $\mathbb{R}^n$, consider the following sets:

$$\mathrm{A}(C) = \{y \in \mathbb{R}^n \mid x + \lambda y \in C,\ \forall x \in C,\ \forall \lambda \geq 0\},$$
$$\mathrm{A}'(C) = \{y \in \mathbb{R}^n \mid \exists x \in C,\ x + \lambda y \in C,\ \forall \lambda \geq 0\},$$
$$\mathrm{A}''(C) = \{y \in \mathbb{R}^n \mid y = \lim_{\nu \to \infty} \lambda_\nu x_\nu,\ (\lambda_\nu)_{\nu \in \mathbb{N}} \subset \mathbb{R},\ \lambda_\nu \downarrow 0,\ (x_\nu)_{\nu \in \mathbb{N}} \subset C\}.$$

a) Show that $\mathrm{A}(C)$ is a convex cone containing the origin which coincides with the set of vectors y such that $C + \{y\} \subset C$.
$\mathrm{A}(C)$ is called *asymptotic cone* (or *recession cone*) of C.
b) Noting that $y = \lim_{k \to \infty}(1/k)(x + ky)$, show that $\mathrm{A}(C) \subset \mathrm{A}'(C) \subset \mathrm{A}''(C)$. If C is closed, show that $\mathrm{A}''(C) \subset \mathrm{A}(C)$. (Hint: If $y = \lim_{\nu \to \infty} x^\nu$, write $x + \lambda y = \lim_{\nu \to \infty}\big((1 - \lambda\lambda_\nu)x + \lambda\lambda_\nu x^\nu\big)$.)
Deduce then that if C is closed, then C is bounded if and only if $\mathrm{A}(C) = \{0\}$.
c) Prove the following assertions:
(i) $\forall z \in \mathbb{R}^n$, $\mathrm{A}(C + \{z\}) = \mathrm{A}(C)$.
(ii) If $0 \in C$, then $\mathrm{A}(C) \subset C$.
(iii) If K is a convex cone containing the origin, then $\mathrm{A}(K) = K$.
(iv) If V is a nonempty affine set, then $\mathrm{A}(V)$ is precisely the linear subspace M parallel to V.
(v) If C is closed and contains the origin, then $\mathrm{A}(C)$ is the greatest convex cone containing the origin and contained in C.
(vi) If C^1 and C^2 are closed and if $C^1 \subset C^2$, then $\mathrm{A}(C^1) \subset \mathrm{A}(C^2)$.
(vii) If C^1 and C^2 are closed and if $C^1 \cap C^2 \neq \emptyset$, then $\mathrm{A}(C^1 \cap C^2) = \mathrm{A}(C^1) \cap \mathrm{A}(C^2)$.
d) Use the previous results to find the asymptotic cones of the following subsets of $\mathbb{R}^2$:
$C^1 = \{(x_1, x_2) \mid x_1 > 0, x_2 > 1/x_1\}$;
$C^2 = \{(x_1, x_2) \mid x_2 \geq x_1^2\}$;
$C^3 = \{(x_1, x_2) \mid x_1^2 + x_2^2 \leq 1\}$;
$C^4 = \{(x_1, x_2) \mid x_1 > 0, x_2 > 0\} \cup \{(0, 0)\}$.

Exercise 1.18. Let $\{a^1, \ldots, a^m\}$ be m points of $\mathbb{R}^n$ and let $b_1, \ldots, b_m$ be m real numbers. Find the asymptotic cone of the set $\{x \in \mathbb{R}^n \mid a^i \cdot x \geq b_i,\ i = 1, \ldots, m\}$.

Exercise 1.19. For a convex subset Y of $\mathbb{R}^n$ and $e \in -\mathbb{R}^n_+$, define $\widehat{Y} := \{y \in Y \mid y \geq e\}$. Show that if Y is convex, closed, contains $-\mathbb{R}^n_+$ and has a bounded intersection with $\mathbb{R}^n_+$, then $\widehat{Y}$ is bounded from above.

Exercise 1.20. Let C^1 and C^2 be two closed convex sets having $\mathrm{A}(C^1)$, $\mathrm{A}(C^2)$ as asymptotic cones. Show that if $\mathrm{A}(C^1) \cap (-\mathrm{A}(C^2)) = \{0\}$, then $(C^1 + C^2)$ is a closed subset of $\mathbb{R}^n$.
More generally, the cones $K^1, K^2, \ldots, K^m$ are said to be *positively semi-independent* if $x^1+x^2+\ldots+x^m = 0$ together with $x^i \in K^i$, $i = 1, \ldots, m$ imply $x^1 = x^2 = \ldots = x^m = 0$. Show that if the convex closed sets $C^1, C^2, \ldots, C^m$ have positively semi-independent asymptotic cones, then the set $(C^1 + C^2 + \ldots + C^m)$ is closed.

Exercise 1.21. Let $(C^i)_{i\in I}$ be an indexed family of closed convex sets with a nonempty intersection. Show that $\mathrm{A}(\bigcap_{i\in I} C^i) = \bigcap_{i\in I} \mathrm{A}(C^i)$.Then, if $\bigcap_{i\in I} \mathrm{A}(C^i) = \{0\}$, the set $\bigcap_{i\in I} C^i$ is bounded.

Exercise 1.22. a) Let C be a closed cone in $\mathbb{R}^2$. Show that $\overline{\mathrm{co}}\, C = \mathrm{co}\, C$.
b) Let in $\mathbb{R}^3$, $C = \{x \in \mathbb{R}^3 \mid (x_1 - |x_3|)^2 + x_2^2 \leq x_3^2\}$. Show that $\mathrm{co}\, C$ is the union of the open half-space defined by $x_1 > 0$ and of the line $x_1 = x_2 = 0$ while $\overline{\mathrm{co}}\, C$ is the closed half-space defined by $x_1 \geq 0$.

Exercise 1.23. Show that the relative interior $\mathrm{ri}(C)$ of a convex cone is a convex cone with same vertex.

Exercise 1.24. If a convex set B is contained in the relative boundary of a nonempty convex set A, show that $\dim B < \dim A$.

Exercise 1.25. Let in $\mathbb{R}^2$, $A = \{0\} \times \mathbb{R}$ and $B = [0,1] \times \mathbb{R}$. Prove that $\mathrm{ri}(A) \not\subset \mathrm{ri}(B)$ though $A \subset B$.

Exercise 1.26. Let A and B be two convex sets with a nonempty intersection. Is the relative interior of $(A \cap B)$ equal to the intersection of the relative interiors of A and B? Is the closure of $(A \cap B)$ equal to the intersection of the closures of A and B? Show that if $\mathrm{ri}(A) \cap \mathrm{ri}(B) \neq \emptyset$, then $\mathrm{ri}(A \cap B) = \mathrm{ri}(A) \cap \mathrm{ri}(B)$.

Exercise 1.27. Let A and B be two convex sets in $\mathbb{R}^n$.
a) $\overline{A}+\overline{B} \subset \overline{A+B}$ (this inclusion does not depend on A and B being convex);
b) $\overline{A+B} = \overline{\mathrm{ri}(A) + \mathrm{ri}(B)}$;
In order to show that

$$\mathrm{ri}(A + B) = \mathrm{ri}(A) + \mathrm{ri}(B),$$

prove that
c) $\mathrm{ri}(A + B) \subset \mathrm{ri}(A) + \mathrm{ri}(B)$;
d) $\mathrm{ri}(A) + \mathrm{ri}(B) \subset \mathrm{ri}(A + B)$.

Exercise 1.28. Prove that a convex subset of $\mathbb{R}^n$ is closed if and only if its intersection with any line is closed.

2. Separation and Polarity

2.1 Separation of convex sets

2.1.1 Definitions

Let $H = \{x \in \mathbb{R}^n \mid p \cdot x = \alpha\}$ be a hyperplane in $\mathbb{R}^n$ ($p \in \mathbb{R}^n \setminus \{0\}$, $\alpha \in \mathbb{R}$). The sets $F = \{x \in \mathbb{R}^n \mid p \cdot x \leq \alpha\}$ and $G = \{x \in \mathbb{R}^n \mid p \cdot x \geq \alpha\}$ are called *closed half-spaces associated with* (or *determined by*) H. Obviously, they are convex and topologically closed as the inverse images of the closed intervals of $\mathbb{R}$: $]-\infty, \alpha]$ and $[\alpha, +\infty[$ respectively by the (continuous) linear functional: $x \to p \cdot x$. The complementary sets $V = \{x \in \mathbb{R}^n \mid p \cdot x > \alpha\}$ and $U = \{x \in \mathbb{R}^n \mid p \cdot x < \alpha\}$ are obviously convex and topologically open; they are called *open half-spaces associated with* (or *determined by*) H.

Since a nonnull linear functional is onto, it should be noticed that F, G, U, V are nonempty. It is easily verified that $F = \overline{U}$, $G = \overline{V}$, so that, in view of the convexity of U and V, $\mathrm{i}(F) = U$, $\mathrm{i}(G) = V$ and H is the common boundary of F, G, U, V. It is also worth noticing that the previous definitions do not depend on the particular equation defining H. If $H = \{x \in \mathbb{R}^n \mid p \cdot x = \alpha\} = \{x \in \mathbb{R}^n \mid p' \cdot x = \alpha'\}$, let $\lambda \in \mathbb{R}$ be such that $p' = \lambda p$ and $\alpha' = \lambda\alpha$ (Proposition 1.1.9). Then the set $\{x \in \mathbb{R}^n \mid p \cdot x \leq \alpha\}$ coincide with one of the sets $\{x \in \mathbb{R}^n \mid p' \cdot x \leq \alpha'\}$ or $\{x \in \mathbb{R}^n \mid p' \cdot x \geq \alpha'\}$ depending on the sign of λ. Likewise, the set $\{x \in \mathbb{R}^n \mid p \cdot x < \alpha\}$ coincides with one of the sets $\{x \in \mathbb{R}^n \mid p' \cdot x < \alpha'\}$ or $\{x \in \mathbb{R}^n \mid p' \cdot x > \alpha'\}$.

The points of a nonempty subset M of $\mathbb{R}^n$ are said to be *on a same side* of the hyperplane H if M is contained in one of the closed half-spaces associated with H. The points of M are said to be *strictly on a same side* of H if M is contained in one of the open half-spaces associated with H.

A hyperplane H is said to *separate* two nonempty subsets A and B of $\mathbb{R}^n$ if A is contained in one of the closed half-spaces associated to H and if B lies in the opposite closed half space. It is said to *strictly separate* A and B if A and B are contained in opposing open half-spaces.

A hyperplane H is said to be a *support hyperplane* for a subset A of $\mathbb{R}^n$ if H contains at least a point of A and if all the points of A are on a same side of H.

Proposition 2.1.1. *Let H be a hyperplane and A be a convex subset of $\mathbb{R}^n$. A necessary and sufficient condition for the points of A to be strictly on a same side of H is that $A \cap H = \emptyset$.*

Proof. The condition is obviously necessary. Let us prove by contraposition that it is sufficient. Assume $A \cap H = \emptyset$ and let $p \cdot x = \alpha$ be an equation of H. If there exist x and y in A such that $p \cdot x < \alpha < p \cdot y$, then $0 < \lambda = (p \cdot y - \alpha)/(p \cdot y - p \cdot x) < 1$ and

$$\lambda x + (1-\lambda)y = \frac{p \cdot y - \alpha}{p \cdot y - p \cdot x} x + \frac{\alpha - p \cdot x}{p \cdot y - p \cdot x} y \in A \cap H$$

which contradicts $A \cap H = \emptyset$. □

Proposition 2.1.2. *Let A be a nonempty compact subset of $\mathbb{R}^n$. For every hyperplane H of $\mathbb{R}^n$, A has a support hyperplane parallel to H.*

Proof. Let $p \cdot x = \alpha$ be an equation of H. The (continuous) linear functional $x \to p \cdot x$ has a minimum and a maximum in A. Thus there exist β and γ such that $\forall x \in A$, $\beta \leq p \cdot x \leq \gamma$ with $p \cdot x^1 = \beta$ for some $x^1 \in A$ and $p \cdot x^2 = \gamma$ for some $x^2 \in A$. If $\beta < \gamma$, A has exactly two support hyperplanes parallel to H; if $\beta = \gamma$, A is contained in the hyperplane of equation $p \cdot x = \beta = \gamma$ which is a support hyperplane of A. □

2.1.2 Projection on a closed convex set

In the Euclidean space $\mathbb{R}^n$, the separation theorems follow in an elementary way from the geometric structure of $\mathbb{R}^n$ (especially through the definition of the Euclidean norm) as well as from its topological properties (in particular, the fact that closed and bounded subsets are compact). This subsection is devoted to considerations originating from von Neumann and Morgenstern (1947).

Proposition 2.1.3. *Let C be a closed convex subset of $\mathbb{R}^n$.*
(i) For $x \in \mathbb{R}^n$, $\pi \in C$, the following properties are equivalent:
a) $\|x - \pi\| = \min_{y \in C} \|x - y\|$;
b) $(x - \pi) \cdot (y - \pi) \leq 0$, $\forall y \in C$.
(ii) For every $x \in \mathbb{R}^n$, there exists a unique point $\pi(x) \in C$ satisfying a) and b); $\pi(x)$ is said to be the projection of x onto C.

Proof. (i): a) implies b). Indeed, let $y \in C$. Since C is convex, $\lambda y + (1 - \lambda)\pi \in C$ holds for every λ such that $0 < \lambda < 1$. It follows from a) that $\|x - \pi\|^2 \leq \|x - (\lambda y + (1-\lambda)\pi)\|^2 = \|(x - \pi) - \lambda(y - \pi)\|^2 = \|x - \pi\|^2 - 2\lambda(x-\pi)\cdot(y-\pi) + \lambda^2\|y-\pi\|^2$ which implies $0 \leq \lambda^2\|y-\pi\|^2 - 2\lambda(x-\pi)\cdot(y-\pi)$. Dividing by λ and letting λ tend to 0, one gets

$$(x - \pi) \cdot (y - \pi) \leq 0.$$

(i): b) implies a). Let again $y \in C$. $(x-\pi)\cdot(y-\pi) = (x-\pi)\cdot(y-x+x-\pi) = \|x-\pi\|^2 + (x-\pi)\cdot(y-x)$. From b), one deduces $\|x-\pi\|^2 \leq (x-\pi)\cdot(x-y)$. Using the Cauchy-Schwarz inequality, one gets $\|x-\pi\|^2 \leq \|x-\pi\|\ \|x-y\|$ and finally $\|x-\pi\| \leq \|x-y\|$.
(ii) For $x \in \mathbb{R}^n$, the existence of π is proved using a), while b) implies its uniqueness. Indeed, let B be a closed ball with center x and a radius such that $B \cap C \neq \emptyset$. The continuous function $y \to \|x-y\|$ has a minimum at a point π of the compact set $B \cap C$. Obviously, π satisfies a). Then if π^1 and π^2 satisfy b), then $(x-\pi^1)\cdot(\pi^2-\pi^1) \leq 0$ and $(x-\pi^2)\cdot(\pi^1-\pi^2) \leq 0$. Combining both inequalities, one gets $\|\pi^2-\pi^1\| \leq 0$, which implies $\pi^2 = \pi^1$. □

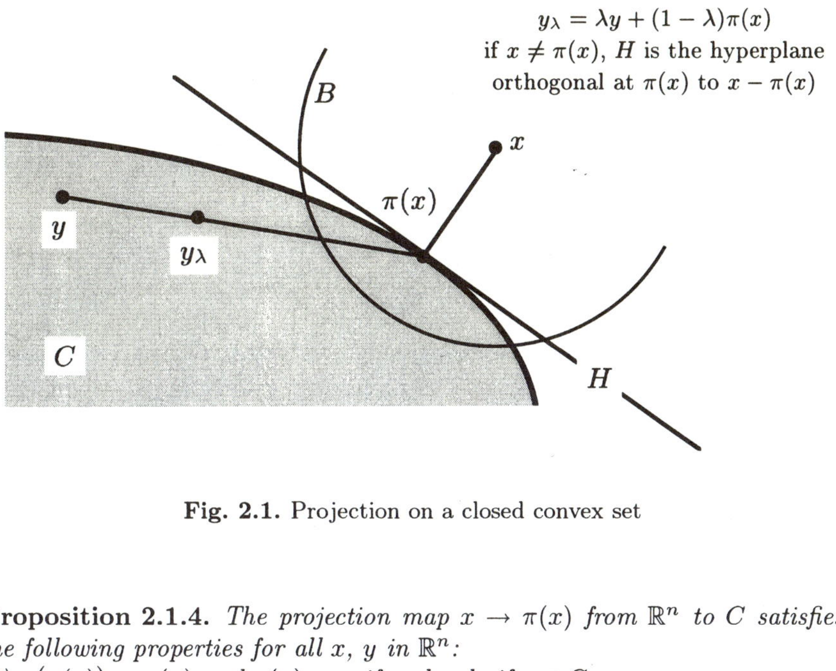

Fig. 2.1. Projection on a closed convex set

Proposition 2.1.4. *The projection map $x \to \pi(x)$ from $\mathbb{R}^n$ to C satisfies the following properties for all x, y in $\mathbb{R}^n$:*
(i) $\pi(\pi(x)) = \pi(x)$ and $\pi(x) = x$ if and only if $x \in C$;
(ii) $(\pi(y)-\pi(x))\cdot(x-y) \leq 0$;
(iii) $\|\pi(x)-\pi(y)\| \leq \|x-y\|$.
In particular, the map $x \to \pi(x)$ is continuous.

Proof. The first part of (i) is obvious, the second part comes from closedness of C.

Applying the property (i) b) of Proposition 2.1.3 respectively to $\pi(x)$ and $\pi(y)$, one gets $(x-\pi(x))\cdot(\pi(y)-\pi(x)) \leq 0$ and $(y-\pi(y))\cdot(\pi(x)-\pi(y)) \leq 0$. Combining the two previous inequalities, one sees that

$$(x-y)\cdot\big(\pi(y)-\pi(x)\big)+\|\pi(x)-\pi(y)\|^2\le 0$$

which proves (ii) and also (iii). Indeed, using the Cauchy-Schwarz inequality,

$$\|\pi(y)-\pi(x)\|^2\le (y-x)\cdot\big(\pi(y)-\pi(x)\big)\le \|y-x\|\ \|\pi(y)-\pi(x)\|.$$

The continuity of the projection map is an immediate consequence of (iii). □

Observe that, as proved in the following proposition, if the closed convex set C is a linear subspace M of $\mathbb{R}^n$, then the projection map π coincides with the projection on M parallel to $M^\perp$ (the orthogonal complement of M).

Proposition 2.1.5. *Let M be a linear subspace of $\mathbb{R}^n$ and π be the projection map onto M. Then for two arbitrary points x and y of $\mathbb{R}^n$:*
(i) $x=\pi(x)+\big(x-\pi(x)\big)$ with $\big(x-\pi(x)\big)\in M^\perp$;
(ii) $\pi(x)\cdot y=x\cdot\pi(y)$ and $\|\pi(x)\|\le\|x\|$.

Proof. From (i) b) in Proposition 2.1.3, one deduces that for every z of M, $\big(x-\pi(x)\big)\cdot\big(z-\pi(x)\big)\le 0$. If y is an arbitrary point of M, applying the previous inequality to $z=y+\pi(x)$ (which belongs to M), one gets: $\big(x-\pi(x)\big)\cdot y\le 0$. Since M is a linear subspace of $\mathbb{R}^n$, it holds that $\big(x-\pi(x)\big)\cdot y=0\ \forall y\in M$, which proves (i).

One then has $\big(x-\pi(x)\big)\cdot\pi(y)=0$. Changing x for y, one has also $\big(y-\pi(y)\big)\cdot\pi(x)=0$. Finally, $x\cdot\pi(y)=\pi(x)\cdot\pi(y)=y\cdot\pi(x)$, which proves the first part of (ii). Since x and y are arbitrary, one has also $\pi(x)\cdot\pi(x)=x\cdot\pi(x)$, which implies $\|\pi(x)\|^2\le\|x\|\ \|\pi(x)\|$ and finally $\|\pi(x)\|\le\|x\|$. It is worth noticing that this last inequality could also be deduced from (iii) in Proposition 2.1.4. □

A last remark is in order. As it follows from the previous proposition, the projection map on a linear subspace is a linear map. Conversely, if the projection map π onto a closed convex subset C of $\mathbb{R}^n$ is a linear map, it follows from $C=\pi(\mathbb{R}^n)$ that C is a linear subspace.

2.1.3 Separation theorems

Proposition 2.1.6. *Given a nonempty closed convex subset C of $\mathbb{R}^n$ and $x^0\notin C$, there exist $p\neq 0$, $p\in\mathbb{R}^n$ and $\alpha\in\mathbb{R}$ such that $p\cdot x\le\alpha$ for all $x\in C$ while $p\cdot x^0>\alpha$.*

Proof. Let $\pi(x^0)$ be the projection of x^0 on C. Define $p=x^0-\pi(x^0)$. By the closedness of C, $\pi(x^0)\neq x^0$, hence $p\neq 0$. It follows from (i) b) in Proposition 2.1.3 that for every $y\in C$, $p\cdot\big(y-\pi(x^0)\big)\le 0$, that is $p\cdot y\le p\cdot\pi(x^0)$. Setting $\alpha=p\cdot\pi(x^0)$, one gets $p\cdot y\le\alpha$ for all $y\in C$. Moreover, $x^0=p+\pi(x^0)$ implies $p\cdot x^0=p\cdot\pi(x^0)+\|p\|^2>\alpha$ since $\|p\|^2>0$.

Notice that the hyperplane $H=\{z\in\mathbb{R}^n\mid p\cdot z=\alpha\}$ is a supporting hyperplane to C at $\pi(x^0)$. □

Corollary 2.1.1. *(Support Theorem) Given a convex subset C of $\mathbb{R}^n$ and y in the boundary of C, there exists $p \neq 0$, $p \in \mathbb{R}^n$ such that $p \cdot x \leq p \cdot y$ for all $x \in \overline{C}$ (in other words, $\overline{C}$ has a supporting hyperplane containing y).*

Proof. Assume first that $\mathrm{i}(C) \neq \emptyset$. Then, $\mathrm{i}(C) = \mathrm{i}(\overline{C})$ (Corollary 1.2.2), so that the boundary of C coincides with the boundary of $\overline{C}$. Hence $y \in \overline{\mathbb{R}^n \setminus \overline{C}}$. Let $(x^{0k})_{k \in \mathbb{N}}$ be a sequence of points of $\mathbb{R}^n \setminus \overline{C}$ converging to y. By the proof of Proposition 2.1.6, for each k there exists $p^k \neq 0$, $p^k \in \mathbb{R}^n$ such that

$$p^k \cdot x \leq p^k \cdot x^{0k} - \|p^k\|^2, \ \forall x \in \overline{C}$$

equivalently, dividing by $\|p^k\|$,

$$\frac{p^k}{\|p^k\|} \cdot x \leq \frac{p^k}{\|p^k\|} \cdot x^{0k} - \|p^k\|, \ \forall x \in \overline{C}.$$

Without loss of generality, one can assume that $(p^k/\|p^k\|)$ tends to $p \neq 0$. Then $\|p^k\| = x^{0k} - \pi(x^{0k})$ tends to 0 since x^{0k} tends to $y \in \overline{C}$ and π is continuous (Proposition 2.1.4). Passing to limits in the previous inequality, one gets

$$p \cdot x \leq p \cdot y, \ \forall x \in \overline{C}.$$

Setting $\alpha = p \cdot y$, the hyperplane $H = \{z \in \mathbb{R}^n \mid p \cdot z = \alpha\}$ is a support hyperplane of $\overline{C}$ containing y.

Assume now that $\mathrm{i}(C) = \emptyset$. Then $\dim C = p < n$ and C is contained in some hyperplane H containing $\operatorname{aff} C$. Let p be a vector orthogonal to H. There exists some α such that $H = \{z \in \mathbb{R}^n \mid p \cdot z = \alpha\}$. Since $C \subset H$, $\overline{C} \subset H$ and H is also a support hyperplane of $\overline{C}$ containing y. □

Corollary 2.1.2. *Given a nonempty convex subset C of $\mathbb{R}^n$ and $y \notin C$, there exists $p \neq 0$, $p \in \mathbb{R}^n$ such that $p \cdot x \leq p \cdot y$ for all $x \in C$.*

Proof. $\overline{C}$ is a closed convex subset of $\mathbb{R}^n$. If $y \notin \overline{C}$, the result is a consequence of Proposition 2.1.6. If $y \in \overline{C}$, y is in the boundary of C and the result is a consequence of Corollary 2.1.1. □

Corollary 2.1.2 allows to prove:

Proposition 2.1.7 (First Separation Theorem). *Let A and B be two nonempty disjoint convex subsets of $\mathbb{R}^n$. Then there exists a hyperplane separating A and B.*

Proof. Define $C = A - B$. The set C is convex, nonempty with $0 \notin C$. In view of Corollary 2.1.2, there exists $p \neq 0$, $p \in \mathbb{R}^n$ such that $p \cdot z \leq 0$ for all $z \in C$. Thus

$$p \cdot x \leq p \cdot y, \ \forall x \in A, \ \forall y \in B.$$

Let then $\alpha = \sup\{p \cdot x \mid x \in A\}$ and $\beta = \inf\{p \cdot y \mid y \in B\}$. One has $\alpha \leq \beta$ and every hyperplane $H = \{z \in \mathbb{R}^n \mid p \cdot z = \gamma\}$ with $\alpha \leq \gamma \leq \beta$ separates A and B. □

Let us remark that the separation is strict if A and B are open. Indeed, each of the sets is contained in the interior of a closed half-space, and thus in the corresponding open half-space.

Proposition 2.1.8 (Second Separation Theorem). *Let A and B be two nonempty disjoint closed convex subsets of $\mathbb{R}^n$. If one of them is compact, then there exists a hyperplane which strictly separates A and B.*

Moreover, there exists some real numbers α and β and some vector p such that

$$\forall a \in A, \forall b \in B, \;\; p \cdot a \leq \alpha < \beta \leq p \cdot b$$

Proof. Assume that A is compact and define $\varepsilon = d(A, B) := \inf\{d(x, y) \mid x \in A, \, y \in B\}$. As shown in Exercise 1.4, $\varepsilon > 0$ so that if $B(0, \varepsilon/2)$ is the open ball centered at the origin with radius $\varepsilon/2$,

$$\big(A + B(0, \frac{\varepsilon}{2})\big) \bigcap \big(B + B(0, \frac{\varepsilon}{2})\big) = \emptyset.$$

The sets $\big(A + B(0, \varepsilon/2)\big)$ and $\big(B + B(0, \varepsilon/2)\big)$ are convex and open. In view of the remark at the end of Proposition 2.1.7, there exist a hyperplane H which strictly separates $\big(A + B(0, \varepsilon/2)\big)$ and $\big(B + B(0, \varepsilon/2)\big)$. A fortiori, H strictly separates A and B.

Finally, let p be a vector orthogonal to H. There exists some β such that $H = \{z \in \mathbb{R}^n \mid p \cdot z = \beta\}$. With no loss of generality, one may assume that $A \subset \{z \in \mathbb{R}^n \mid p \cdot z < \beta\}$ and $B \subset \{z \in \mathbb{R}^n \mid p \cdot z > \beta\}$. If we define $\alpha := \sup\{p \cdot a \mid a \in A\}$, by a classical compactness argument, $\alpha < \beta$ and the conclusion holds. □

Proposition 2.1.6 has as an immediate corollary:

Corollary 2.1.3. *In $\mathbb{R}^n$, a closed convex set C is the intersection of all the closed half-spaces which contain it.*

Proof. The result is trivial if $C = \emptyset$ or if $C = \mathbb{R}^n$. If not, we have only to prove that the intersection of all the closed half-spaces containing C is contained in C. Indeed if $a \notin C$, there exists a hyperplane which strictly separates the compact convex set $\{a\}$ and the closed convex set C. This hyperplane determines two closed half-spaces, one of them contains C and does not contain a. □

Finally, let us remark that Proposition 2.1.6 could now be deduced from the second separation theorem (the sets $\{x^0\}$ and C are respectively a compact convex set and a nonempty closed convex set with an empty intersection). Hence Proposition 2.1.6 and the two separation theorems can be thought of as logically equivalent versions of a same *separation principle.* This circularity and the particular form of the first separation theorem (where no topological assumption is made on the convex sets to be separated) characterize the

Euclidean space $\mathbb{R}^n$, and depend heavily on its algebraic and topological properties.

Moreover, using the properties of $\mathbb{R}^n$, Corollary 2.1.1 can be strengthened as follows:

Proposition 2.1.9. *Given a convex subset C of $\mathbb{R}^n$ and y in the relative boundary of C, there exists a supporting hyperplane H to $\overline{C}$, passing through y, which does not contain C and such that one of the two open half-spaces determined by H contains all of* $\mathrm{ri}(C)$ *(H is said to be a nontrivial supporting hyperplane to C and the points of $H \cap C$ are called proper support points of C).*
Moreover, if F is the linear subspace of $\mathbb{R}^n$ parallel to $\operatorname{aff} C$*, there exists $p \in F \setminus \{0\}$ such that $p \cdot x = p \cdot y$ is the equation of H.*

Proof. If $\mathrm{i}(C) \neq \emptyset$, then $\dim C = n$ and H, the existence of which follows from Corollary 2.1.1, cannot contain C. If $\mathrm{i}(C) = \emptyset$, then $\dim C < n$. Let $m \in \operatorname{aff} C$. Then $(C - \{m\})$ is a convex set contained in the linear subspace $F = (\operatorname{aff} C - \{m\})$. In this finite dimensional linear space, the interior of $(C - \{m\})$ is precisely equal to $(\mathrm{ri}(C) - \{m\})$ and $y - m$ is in the boundary of $(C - \{m\})$. In view of Corollary 2.1.1 applied in this finite dimensional space, there exists a linear functional $f' : (\operatorname{aff} C - \{m\}) \to \mathbb{R}$ such that $f'(z) \leq f'(y-m)$, $\forall z \in \overline{(C - \{m\})}$ i.e. $f'(z) \leq f'(y-m)$, $\forall z \in (\overline{C} - \{m\})$. Moreover, in this finite dimensional space, $f'(z) < f'(y-m)$, $\forall z \in \mathrm{i}(C - \{m\})$ i.e. $f'(z) < f'(y-m)$, $\forall z \in \mathrm{ri}(C) - \{m\}$. Let now f be the linear functional on $\mathbb{R}^n$ defined by

$$f(x) = \begin{cases} f'(x) & \text{if } x \in F; \\ 0 & \text{if } x \in F^\perp. \end{cases}$$

Equivalently, $f(x) = (f' \circ \pi)(x)$, where π is the orthogonal projection on F. Obviously, f extends f'. We then have: $f(z) \leq f(y-m)$, $\forall z \in (\overline{C} - \{m\})$ while $f(z) < f(y-m)$, $\forall z \in \mathrm{ri}(C) - \{m\}$, thus $f(x) \leq f(y)$, $\forall x \in \overline{C}$, $f(x) < f(y)$, $\forall x \in \mathrm{ri}(C)$. The hyperplane of equation $f(x) = f(y)$ is a supporting hyperplane to $\overline{C}$, passing through y, which does not contain C. Let $p \in \mathbb{R}^n$ be such that $f(x) = p \cdot x$, $\forall x \in \mathbb{R}^n$. Since, by definition of f, $f(x) = p \cdot x = 0$, $\forall x \in F^\perp$, obviously $p \in F^{\perp\perp} = F$. □

Corollary 2.1.4. *Let C be a convex subset of $\mathbb{R}^n$ and let y be a point in $F \setminus \mathrm{ri}(C)$. Then there exists $p \in F \setminus \{0\}$ such that $p \cdot x < p \cdot y$ for all $x \in \mathrm{ri}(C)$.*

Proof. If $y \notin \overline{C}$, the existence of p follows from the second separation theorem. If $y \in \overline{C}$, the existence of p follows from proposition 2.1.9. Indeed, let $p \in F \setminus \{0\}$ and $z \in C$ be such that $p \cdot x \leq p \cdot y$ $\forall x \in C$, $p \cdot z < p \cdot y$. If there exists $x \in \mathrm{ri}(C)$ such that $p \cdot x = p \cdot y$, then $p \cdot u > p \cdot y$ for some $u = x - \lambda(z - x) \in C$ with $\lambda > 0$, which contradicts the definition of p. □

Finally, Corollary 2.1.4 allows for the following strengthening of the first separation theorem:

Proposition 2.1.10. *Let A and B be two nonempty convex subsets of $\mathbb{R}^n$. In order that there exist a hyperplane separating A and B properly (that is not containing both A and B), it is necessary and sufficient that* $\mathrm{ri}(A) \cap \mathrm{ri}(B) = \emptyset$.

Proof. Assume first that $\mathrm{ri}(A) \cap \mathrm{ri}(B) = \emptyset$. As proved in Exercise 1.27, $0 \notin (\mathrm{ri}(B) - \mathrm{ri}(A)) = \mathrm{ri}(B - A)$. By the previous corollary, there exists $p \in \mathbb{R}^n \setminus \{0\}$ such that $p \cdot z > 0$ for all $z \in \mathrm{ri}(B - A)$, i.e., $p \cdot x < p \cdot y$ for all $x \in \mathrm{ri}(A)$, $y \in \mathrm{ri}(B)$. Define $\alpha = \sup\{p \cdot x \mid x \in A\}$ and $\beta = \inf\{p \cdot y \mid y \in B\}$. Obviously $\alpha \leq \beta$. Let γ be such that $\alpha \leq \gamma \leq \beta$. The hyperplane of equation $p \cdot z = \gamma$ cannot contain both A and B since it would contradict the above strict inequality written for the points of $\mathrm{ri}(A)$ and $\mathrm{ri}(B)$ respectively.
Let us assume conversely that a hyperplane of equation $p \cdot z = \gamma$ separates A and B (for example, $p \cdot x \leq \gamma \leq p \cdot y$ for all $x \in A$, $y \in B$) and does not contain B. It is easily seen that $p \cdot y > \gamma$, $\forall y \in \mathrm{ri}(B)$, which proves that $\mathrm{ri}(A) \cap \mathrm{ri}(B) = \emptyset$. □

2.2 Polars of convex sets and orthogonal subspaces

For $A \subset \mathbb{R}^n$, the set

$$A^\circ := \{y \in \mathbb{R}^n \mid x \cdot y \leq 1 \ \forall x \in A\}$$

is called the *polar* of A. The polar of A° is

$$A^{\circ\circ} := (A^\circ)^\circ = \{x \in \mathbb{R}^n \mid y \cdot x \leq 1 \ \forall y \in A^\circ\}$$

If $A = \{a\}$ for some $a \neq 0$, then $\{a\}^\circ = \{y \in \mathbb{R}^n \mid a \cdot y \leq 1\}$ is a closed half-space. If A contains more than one point, $A = \bigcap_{x \in A} \{y \in \mathbb{R}^n \mid x \cdot y \leq 1\}$ is an intersection of closed half-spaces.

If $A \subset B$, it is easily verified that $B^\circ \subset A^\circ$. Moreover, $(\mathrm{co}\, A)^\circ = A^\circ$ and $A \subset B \subset \overline{A}$ implies $B^\circ = A^\circ$.

Proposition 2.2.1. *(i) For every subset A of $\mathbb{R}^n$, the set A° is convex, closed and contains the origin.*
(ii) If A is the ball $\overline{B(0,r)}$, then A° is the closed ball $\overline{B(0,1/r)}$.
(iii) If A is a cone with vertex 0, A° is the closed convex cone with vertex 0, called polar cone of A

$$A^\circ = \{y \in \mathbb{R}^n \mid x \cdot y \leq 0 \ \forall x \in A\}.$$

(iv) If A is a linear subspace of $\mathbb{R}^n$, A° is a linear subspace of $\mathbb{R}^n$, the orthogonal complement of A:

$$A^\circ = A^\perp = \{y \in \mathbb{R}^n \mid x \cdot y = 0 \ \forall x \in A\}.$$

Proof. (i) Immediate.

(ii) Let $y \in \overline{B(0,1/r)}$. For every $x \in \overline{B(0,r)}$, one has $x \cdot y \leq \|x\|\,\|y\| \leq 1$, which proves that $y \in A^\circ$. Conversely, assume $y \in A^\circ$. Then $r\underline{(y/\|y\|)} \in A$ and $r(y/\|y\|) \cdot y \leq 1$, which proves that $\|y\| \leq 1/r$ and that $y \in \overline{B(0,1/r)}$.

(iii) It suffices to prove that $A^\circ \subset \{y \in \mathbb{R}^n \mid x \cdot y \leq 0\ \forall x \in A\}$. Let $y \in A^\circ$. For every $x \in A$, since A is a cone, one has $\lambda x \cdot y \leq 1$, $\forall \lambda > 0$ which implies $x \cdot y \leq 1/\lambda$, $\forall \lambda > 0$. Passing to limit when λ tends to $+\infty$, one gets $x \cdot y \leq 0$ for every $x \in A$.

(iv) Since A is a cone, it follows from (iii) that $A^\circ = \{y \in \mathbb{R}^n \mid x \cdot y \leq 0\ \forall x \in A\}$. Remarking that if $x \in A$, then $-x \in A$, (iv) is straightforward. □

Proposition 2.2.2. *Let A be a subset of $\mathbb{R}^n$. In order that $A = A^{\circ\circ}$ it is necessary and sufficient that A be closed convex and contain the origin.*

Proof. If $A = A^{\circ\circ}$, it follows from Proposition 2.2.1 (i) that A is closed, convex and contains 0.

Now let A be closed, convex and containing 0. By definition of $A^{\circ\circ}$, $A \subset A^{\circ\circ}$. We prove by contraposition the converse inclusion. Assume $x \in A^{\circ\circ}$, $x \notin A$. By the second separation theorem, there exist $y \in \mathbb{R}^n \setminus \{0\}$ and $\alpha \in \mathbb{R}$ such that $y \cdot z < \alpha < y \cdot x$, $\forall z \in A$. Since $0 \in A$, $\alpha > 0$ and $(1/\alpha)y \cdot z < 1 < (1/\alpha)y \cdot x$, $\forall z \in A$. Thus $(1/\alpha)y \in A^\circ$ and $x \notin A^{\circ\circ}$, a contradiction. □

Proposition 2.2.3. *Let A be a closed convex subset of $\mathbb{R}^n$. Then A is compact if and only if $0 \in \mathrm{i}(A^\circ)$. Dually, if $0 \in A$, then A° is compact if and only if $0 \in \mathrm{i}(A)$.*

Proof. If A is compact, let $r > 0$ be such that $A \subset \overline{B(0,r)}$. Then

$$A^\circ \supset B^\circ = \overline{B(0,1/r)} \supset B_0(0,1/r),$$

which proves that $0 \in \mathrm{i}(A^\circ)$. Conversely, if $0 \in \mathrm{i}(A^\circ)$, let $r > 0$ be such that $B_0(0,r) \subset A^\circ$. Then

$$A \subset A^{\circ\circ} \subset \big(B_0(0,r)\big)^\circ = \overline{B(0,1/r)};$$

A is bounded, thus compact.

If $0 \in A$, then $A = A^{\circ\circ}$. Applying the first part of the proposition, one gets: A° is compact if and only if $0 \in \mathrm{i}(A^{\circ\circ}) = \mathrm{i}(A)$. □

2.3 Application to Minkowski-Farkas' lemma

2.3.1 Minkowski-Farkas' lemma

For a $(m \times n)$-matrix A with $a^1, \ldots, a^n$ in $\mathbb{R}^m$ as column-vectors, let us consider the sets

$$B = \{p \in \mathbb{R}^m \mid {}^t pA \leq 0\} = \{p \in \mathbb{R}^m \mid p \cdot a^i \leq 0,\ i = 1, \dots, n\},$$

the set of solutions to a system of n homogeneous linear inequalities, and

$$C = \{z \in \mathbb{R}^m \mid z = A\lambda,\ \lambda \in \mathbb{R}^n_+\} = \{z \in \mathbb{R}^m \mid z = \sum_{i=1}^{n} \lambda_i a^i,\ \lambda = (\lambda_i) \in \mathbb{R}^n_+\}$$

the convex cone (with vertex 0) containing the origin, generated in $\mathbb{R}^m$ by the finite system of points $\{a^1, \dots, a^n\}$. One obviously has $C^\circ = B$. In order to prove that $B^\circ = C$, we first establish in Lemmas 2.3.1, 2.3.2, 2.3.3 a result interesting by itself: the closedness of the convex cone, containing the origin, generated by a finite system of points. One will find an alternate proof of this auxiliary result in the exercise 2.18 at the end of this chapter.

Lemma 2.3.1. *If $(a^i)_{i\in I}$, $I \subset \{1, \dots, n\}$, is a system of linearly independent vectors of $\mathbb{R}^m$, then the convex cone (with vertex 0), containing the origin, generated by the a^i*

$$C_I = \{z \in \mathbb{R}^m \mid z = \sum_{i\in I} \lambda_i a^i,\ \lambda_i \geq 0\ \forall i \in I\}$$

is closed.

Proof. Indeed, let $(z^\nu)_{\nu\in\mathbb{N}}$ be a sequence of points of C_I converging to z. We can write $z^\nu = \sum_{i\in I} \lambda_i^\nu a^i$, $\lambda_i^\nu \geq 0\ \forall i \in I$. If all the λ_i^ν are bounded, there exists a strictly increasing sequence of integers $(\nu_k)_{k\in\mathbb{N}}$ such that $\lambda_i^{\nu_k} \to \lambda_i$, $\forall i \in I$. Then $z = \sum_{i\in I} \lambda_i a^i$ with $\lambda_i \geq 0$, $\forall i \in I$, which proves that $z \in C_I$.

Hence we have only to prove (and we will prove by contraposition) that all the λ_i^ν are bounded. If it is not the case, let $(\nu_k)_{k\in\mathbb{N}}$ be an increasing sequence of integers such that

$$\sup_{i\in I} \lambda_i^{\nu_k} \to +\infty \text{ and } \frac{\lambda_i^{\nu_k}}{\sup_{i\in I} \lambda_i^{\nu_k}} \to \lambda_i.$$

Let us remark that at least one of the λ_i is not equal to 0. Then dividing by $\sup_{i\in I} \lambda_i^{\nu_k}$ both sides of the equality

$$z^{\nu_k} = \sum_{i\in I} \lambda_i^{\nu_k} a^i$$

and passing to limit, we get $0 = \sum_{i\in I} \lambda_i a^i$, which contradicts the linear independence of vectors a^i. □

Lemma 2.3.2. *For every $z \in C$, there exists $I_z \subset \{1, \dots, n\}$ such that the system $(a_i)_{i\in I_z}$ is linearly independent and $z = \sum_{i\in I_z} \lambda_i a^i$.*

Proof. Let $z \in C$ and define

$$d = \inf\{\operatorname{card} I \mid I \subset \{1, \ldots, n\},\ z = \sum_{i \in I} \lambda_i a^i,\ \lambda_i > 0\ \forall i \in I\}$$

where $\operatorname{card} I$ denotes the number of elements of I. Let us show that if $\operatorname{card} I = d$, then the system $(a_i)_{i \in I}$ is linearly independent. Indeed if not, let $\sum_{i \in I} \mu_i a^i = 0$ be a relation where some real number μ_i is not equal to 0 (without loss of generality, we can assume that some μ_i is < 0). As in the proof of Caratheodory's theorem (Proposition 1.1.2), we can write

$$z = \sum_{i \in I} \lambda_i a^i + \beta \sum_{i \in I} \mu_i a^i = \sum_{i \in I} (\lambda_i + \beta \mu_i) a^i$$

with

$$\beta = \inf_{\{i \mid \mu_i < 0\}} \left(\frac{-\lambda_i}{\mu_i}\right).$$

Note that $\beta > 0$, $\lambda_i + \beta\mu_i \geq 0$, $\forall i \in I$ and that for some i, $\lambda_i + \beta\mu_i = 0$. Thus we have expressed z as a linear combination with strictly positive coefficients of less than d elements of the system $(a^i)_{i \in I}$, a contradiction with the choice of I. □

Lemma 2.3.3. *C is closed.*

Proof. Let $\mathcal{F} = \{I \subset \{1, \ldots, n\} \mid (a_i)_{i \in I}$ is linearly independent $\}$. The set $\mathcal{F}$ is finite and it follows from Lemma 2.3.2 that $C = \bigcup_{I \in \mathcal{F}} C_I$. It follows from Lemma 2.3.1 that each C_I is closed; hence, C is closed. □

Corollary 2.3.1. $B^\circ = C$.

Proof. We already know that $C^\circ = B$. Since C is convex, closed (Lemma 2.3.3) and contains the origin, it follows from Proposition 2.2.2 that $C = C^{\circ\circ} = B^\circ$. □

We can now state and prove Minkowski-Farkas' lemma (which is sometimes referred as Farkas' lemma):

Proposition 2.3.1 (Minkowski-Farkas' lemma). *Let $f^1, \ldots, f^n$ and g be linear functionals defined on $\mathbb{R}^m$. In order that the following implication holds:*

$$f^k(x) \leq 0\ \forall k = 1, \ldots, n \Rightarrow g(x) \leq 0$$

it is necessary and sufficient that there exist $\lambda_1, \ldots \lambda_n \in \mathbb{R}_+$ such that $g = \sum_{k=1}^{n} \lambda_k f^k$.

Proof. Let $a^1, \ldots, a^n, b$ in $\mathbb{R}^m$ be such that $f^k(p) = p \cdot a^k$, $\forall k = 1, \ldots, n$ and $g(p) = p \cdot b$. Consider the sets B and C defined above. The equality $B^\circ = C$ shows that the condition is necessary. On the other hand, the condition is obviously sufficient. □

2.3.2 And its corollaries

Minkowski-Farkas' Lemma, fundamental for the duality theory in Linear Programming, has various corollaries. The first one is simply a translation in terms of matrices of Proposition 2.3.1.

Corollary 2.3.2. *Let A be a $(m \times n)$-matrix and $b \in \mathbb{R}^m$. The equation $Ax = b$ has a solution $x \geq 0$ in $\mathbb{R}^n$ if and only if the following implication is true*

$$ {}^t pA \leq 0, \ p \in \mathbb{R}^m \ \Rightarrow \ p \cdot b \leq 0. $$

Corollary 2.3.3. *Let $f^1, \ldots, f^q$, $h^1, \ldots, h^r$ and g be linear functionals defined on $\mathbb{R}^m$. In order that the following implication holds:*

$$ \left. \begin{array}{ll} f^k(x) = 0 & \forall k = 1, \ldots, q \\ h^k(x) \leq 0 & \forall k = 1, \ldots, r \end{array} \right\} \Rightarrow \ g(x) \leq 0 $$

it is necessary and sufficient that there exist $\mu_1, \ldots, \mu_q$ in $\mathbb{R}$, $\nu_1, \ldots, \nu_r$ in $\mathbb{R}_+$ such that $g = \sum_{k=1}^{q} \mu_k f^k + \sum_{k=1}^{r} \nu_k h^k$.

Proof. Apply Proposition 2.3.1 to the linear functionals f^k, $k = 1, \ldots, q$, $-f^k$, $k = 1, \ldots, q$, h^k, $k = 1, \ldots, r$ and g. □

Corollary 2.3.4. *Let A be a $(m \times n)$-matrix and $b \in \mathbb{R}^m$. In order that the following implication*

$$ {}^t pA \leq 0, \ p \in \mathbb{R}^m_+ \ \Rightarrow \ p \cdot b \leq 0 $$

holds true, it is necessary and sufficient that there exist $x \in \mathbb{R}^n_+$ such that $Ax \geq b$.

Proof. Apply Corollary 2.3.2 to the matrix $B = (\, A \quad -I \,)$, where I is the identity matrix of $\mathbb{R}^m$. □

Corollary 2.3.5. *Let A be a $(m \times n)$-matrix and $b \in \mathbb{R}^m$. In order that the following implication*

$$ {}^t pA = 0, \ p \in \mathbb{R}^m_+ \ \Rightarrow \ p \cdot b \leq 0 $$

holds true, it is necessary and sufficient that there exist $x \in \mathbb{R}^n$ such that $Ax \geq b$.

Proof. Apply Corollary 2.3.2 to the matrix $B = (\, A \quad -A \quad -I \,)$. □

Note that the previous results can be stated in the form of two alternatives. For instance, an equivalent statement of the conclusion of Corollary 2.3.4 is that one and only one of the following alternatives holds:

either there exists $x \in \mathbb{R}^n_+$ such that $Ax \geq b$
or there exists $p \in \mathbb{R}^m_+$ such that ${}^t pA \leq 0$ and $p \cdot b > 0$.
Other pairs of alternatives can be found using Minkowski-Farkas' lemma. Some ones are dealt with in Exercises 2.17, 2.18. The previous pair is used in the proof of the following corollary.

Corollary 2.3.6. *Let K be a skew-symmetric $(n \times n)$-matrix $({}^t K = -K)$. Then the system of linear inequalities with a nonnegativity constraint $Kx \geq 0$, $x \geq 0$ has a solution verifying $x + Kx \gg 0$.*

Proof. Let $i \in \{1, \ldots, n\}$ and e^i be the ith vector of the natural basis of $\mathbb{R}^n$. In view of the alternate statement of Corollary 2.3.4 applied to the skew-symmetric matrix K, it follows that one and only one of the two following systems:

$$\begin{cases} Kx & \geq & e^i \\ x & \geq & 0 \end{cases}, \quad \begin{cases} Kx & \geq & 0 \\ x & \geq & 0 \\ x \cdot e^i & > & 0 \end{cases}$$

has a solution. Let us denote by $x^i \in \mathbb{R}^n$ the solution of one of these systems. Thus for every i, one can find $x^i \geq 0$ such that $Kx^i \geq 0$ and $(x^i + Kx^i)_i > 0$. Setting $x = x^1 + \ldots + x^n$, we see that $x \geq 0$, $Kx \geq 0$, and $(x + Kx)_i \geq (x^i + Kx^i)_i > 0$, $\forall i = 1, \ldots, n$, that is $x + Kx \gg 0$. □

Exercises

Exercise 2.1. If an hyperplane H separates two convex sets A and B, one says that H separates A and B *properly* if A and B are not both contained in H. The hyperplane H separates A and B *nicely* if in addition one of the sets $A \cap H$ or $B \cap H$ is empty. If both $A \cap H$ and $B \cap H$ are empty, H strictly separates A and B. Finally, following Rockafellar, one says that H *strongly* separates A and B if H lies strictly between two translates of H that separate A and B. Clearly, strong separation $\Rightarrow$ strict separation $\Rightarrow$ nice separation $\Rightarrow$ proper separation. Give examples in $\mathbb{R}^2$ showing that none of these implications can be reversed.

Exercise 2.2. Let A and B be two nonempty convex sets verifying: $A \cap B = \emptyset$, A is closed, $A \cup B = \mathbb{R}^n$. Show that A is a closed half-space.

Exercise 2.3. Let A be a closed set with a nonempty interior. If through each point of its boundary passes a supporting hyperplane, show that A is convex. What if the assumption $\mathrm{i}(A) \neq \emptyset$ is removed?

Exercise 2.4. For every $x \neq 0$ in $\mathbb{R}^n$, show that $\bigcup_{n \in \mathbb{N}} B(nx, n\|x\|)$ (the union of all open balls with center nx and radius $n\|x\|$) is a half-space.

Exercise 2.5. Let C be a nonempty cone (with vertex 0). If C is contained in the open half-space $\{x \in \mathbb{R}^n \mid p \cdot x < \alpha\}$, show that $\alpha \geq 0$ and that C is contained in the closed half-space $\{x \in \mathbb{R}^n \mid p \cdot x \leq 0\}$.

Exercise 2.6. Let K be a nonempty closed convex cone (with vertex 0) not equal to $\mathbb{R}^n$. Show that there exists $p \neq 0$ in $\mathbb{R}^n$ such that $p \cdot x \leq 0$ for every $x \in K$. Then show that K is the intersection of the homogeneous closed half-spaces which contain it (calling homogeneous closed half-space one with the origin on its boundary).

Exercise 2.7. Let C be a convex subset of $\mathbb{R}^n$ verifying $C \cap \mathbb{R}^n_{++} = \emptyset$. Show that there exists $p \in \mathbb{R}^n_+ \setminus \{0\}$ such that $\sup\{p \cdot x \mid x \in C\} \leq 0$.

Exercise 2.8. Using the exercise 1.2.3, prove the following result which extends the second separation theorem: If C^1 and C^2 are nonempty disjoint closed convex sets in $\mathbb{R}^n$ such that $\mathrm{A}(C^1) \cap \mathrm{A}(C^2) = \{0\}$, then there exists a hyperplane H separating C^1 and C^2 strongly.

Exercise 2.9. Let C be a closed convex subset of $\mathbb{R}^n$.
a) If x^1 and x^2 belong to the boundary of C and if the open line segment $]x^1, x^2[$ has a nonempty intersection with the boundary of C, show that the whole closed line segment $[x^1, x^2]$ is contained in the boundary of C.
b) Assume now that for every $p \in \mathbb{R}^n \setminus \{0\}$, the set $\{x \in C \mid p \cdot x \geq p \cdot x', \ \forall x' \in C\}$ contains at most one element. Show that if C is not a singleton, then $\mathrm{i}(C) \neq \emptyset$. Using a), deduce that C is strictly convex (that is, $\lambda x^1 + (1-\lambda)x^2 \in \mathrm{i}(C)$ holds true for every $(x^1, x^2) \in C \times C$ such that $x^1 \neq x^2$ and for every $\lambda \in]0, 1[$).
c) Observe conversely that if $\dim C = n$ and C is strictly convex, then for every $p \in \mathbb{R}^n \setminus \{0\}$, the set $\{x \in C \mid p \cdot x \geq p \cdot x', \ \forall x' \in C\}$ contains at most one element.

Exercise 2.10. Determine the polars of the following sets:

$$C^1 = \{x \in \mathbb{R}^n \mid x_k \geq 0, \ k = 1, \ldots, n \text{ and } \textstyle\sum_{k=1}^n x_k \leq 1\}$$
$$C^2 = \{x \in \mathbb{R}^n \mid \textstyle\sum_{k=1}^n |x_k| \leq 1\}$$
$$C^3 = \{x \in \mathbb{R}^2 \mid (x_1 - 1)^2 + x_2^2 \leq 1\}$$
$$C^4 = \{x \in \mathbb{R}^n \mid x_1 \leq 1 - \sqrt{1 + x_2^2}\}.$$

Exercise 2.11. Let C be a closed convex cone in $\mathbb{R}^n$ and let C° be the polar cone of C.
a) If C is a linear subspace of $\mathbb{R}^n$, show that $C = C \cap (-C)$ and that $(-C^\circ) \cap C = \{0\}$.
b) For $x \in C$, defining $\pi(x)$ as the projection of x on $(-C)$, show that $x - \pi(x) \in (-C)^\circ \cap C$.
c) Conclude that in order that a closed convex cone is a linear subspace of $\mathbb{R}^n$ it is necessary and sufficient that $(-C^\circ) \cap C = \{0\}$.

Exercise 2.12. If C and K are convex cones, $(C + K)^\circ = C^\circ \cap K^\circ$. If, moreover, C and K are closed, $(C \cap K)^\circ = \overline{C^\circ + K^\circ}$.
Conclude that if C is a convex cone in $\mathbb{R}^n$, then $\dim C + \dim(C^\circ \cap -C^\circ) = n$ and $\dim C^\circ + \dim(C \cap -C) \leq n$ (see the exercise 1.15). Give an example where this inequality is strict.

Exercise 2.13. For a closed convex cone C in $\mathbb{R}^n$, show that the three following properties are equivalent:
a) $C \cap (-C) = \{0\}$;
b) C has a nonempty convex subset B with $0 \notin B$ such that every $x \in C$, $x \neq 0$ has a unique representation of the form λb, where $b \in B$ and $\lambda > 0$;
c) $\mathrm{i}(C^\circ) \neq \emptyset$ (where C° is the polar cone of C).
Hint: Show that c) $\Rightarrow$ b) $\Rightarrow$ a) $\Rightarrow$ c).

Exercise 2.14. In $\mathbb{R}^n$, let C be a closed convex cone, C° be the polar cone of C and K be a convex compact set. Show that $K \cap C^\circ \neq \emptyset$ if and only if the following holds: $\forall p \in C,\ \exists z \in K,\ p \cdot z \leq 0$.

Exercise 2.15. In $\mathbb{R}^n$, let C be a closed convex cone, C° be the polar cone of C. In order that $C \cap \mathbb{R}^n_+ = \{0\}$ show that it is necessary and sufficient condition that C° contain an element $p \gg 0$. What if C is a linear subspace of $\mathbb{R}^n$?

Exercise 2.16 (Alternative Theorem). Let $\Delta^{n-1} = \{x \in \mathbb{R}^n_+ \mid \sum_{k=1}^n x_k = 1\}$ denote the unit-simplex of $\mathbb{R}^n$. Given a $(m \times n)$-matrix A, consider the set $Z = \{z \in \mathbb{R}^m \mid z = Ax, x \in \Delta^{n-1}\}$. If $a^1, \ldots, a^n$ denote the column-vectors of A, show that Z is the convex hull of the points $a^1, \ldots, a^n$. Deduce that Z is convex and compact.
Separating Z from $\mathbb{R}^m_+$, show that one and only one of the following alternatives holds true:
either $\exists x \in \Delta^{n-1},\ Ax \geq 0$
or $\exists p \in \Delta^{m-1},\ {}^tpA \ll 0$.

The Alternative theorem has several variants. The following one is based on the exercise 2.15.

Exercise 2.17 (Stiemke's lemma). Given a $(m \times n)$-matrix A, consider the set $H = \{z \in \mathbb{R}^n \mid {}^tz = {}^tpA,\ p \in \mathbb{R}^m\}$.
Show that H is a linear subspace of $\mathbb{R}^n$. Conclude that one and only one of the following alternatives holds true:
either $\exists x \in \mathbb{R}^n,\ x \gg 0,\ Ax = 0$
or $\exists p \in \mathbb{R}^m,\ {}^tpA > 0$.

Exercise 2.18. The aim of the last exercise is to give an alternate proof of the result stated in Lemma 2.3.3: the convex cone, containing the origin, generated by a finite system of points is closed.
a) Let A be a compact convex subset of $\mathbb{R}^n$ which does not contain the origin. Show that the pointed convex cone generated by A, $C = \bigcup_{\lambda \geq 0} \lambda A$, is closed.
b) Let C^1 and C^2 be two closed convex cones such that $C^1 \cap (-C^2) = \{0\}$. Show that $C := C^1 + C^2$ is a closed convex cone (see the exercise 1.2.3).
c) Consider the pointed convex cone C generated by the finite system of

points $\{a^1, \ldots, a^p\}$ and denote by A the convex hull of $\{a^1, \ldots, a^p\}$. We will distinguish two cases.

If $0 \notin A$, show that C is closed.

If $0 \in A$, show that $C \cap (-C)$ contains some point of $\{a^1, \ldots, a^p\}$. Hence, without loss of generality, one can assume:
$\{a^1, \ldots, a^p\} = \{a^1, \ldots, a^q\} \bigcup \{a^{q+1}, \ldots, a^p\}$ with
$a^1, \ldots, a^q \notin C \cap (-C)$ and $a^{q+1}, \ldots, a^p \in C \cap (-C)$.
Let us denote by C^1 the pointed convex cone generated by $\{a^1, \ldots, a^q\}$ and by C^2 the pointed convex cone generated by $\{a^{q+1}, \ldots, a^p\}$. Then show that
(i) $0 \notin \text{co}\{a^1, \ldots, a^q\}$, which implies that C^1 is closed
(ii) $C^2 = C \cap (-C)$, which implies that C^2 is closed.
d) Using the previous results, prove that $C = C^1 + C^2$ is closed.

3. Extremal Structure of Convex Sets

3.1 Extreme points and faces of convex sets

3.1.1 Definitions

Let A be a convex subset of $\mathbb{R}^n$. A point $x \in A$ is an *extreme point* of A if and only if there exists no open segment contained in A and containing x, that is, if and only if there is no way to express x as a convex combination $x = \lambda y + (1-\lambda)z$ such that $y \in A$, $z \in A$ and $0 < \lambda < 1$, except by taking $x = y = z$. Easy computations show that an equivalent characterization for an extreme point x of A is that the relations $x = (1/2)y + (1/2)z$ together with $y \in A$, $z \in A$ imply $x = y = z$. Obviously, if $A = \{a\}$, then a is an extreme point of A.

An *extreme subset* or *face* of A is a *convex* subset F of A such that every closed segment of A with a relative interior point in F has both endpoints in F, that is, a convex subset F of A verifying:

$$\left.\begin{array}{l} x \in F \cap \,]y,z[\\ y \in A,\ \ z \in A \end{array}\right\} \Rightarrow y \in F, z \in F.$$

The empty set $\emptyset$ and A are faces of A. The extreme points of A are the 0-dimensional faces of A. If F is a half-line face of A, F is called *extreme ray* and the direction of F is called *extreme direction* of A.
Exercise 3.1 collects a variety of elementary but useful properties of the faces and of the facial structure of a convex set. In particular, every intersection of faces of A is a face of A. If F is a face of A, every face of F is a face of A. Every distinct from A and nonempty face F of A is contained in the relative boundary of A. Thus, as shown in the exercise 1.24, $\dim F < \dim A$. In particular when A contains more than one point, every extreme point of A belongs to the relative boundary of A. Finally, last but not least property, the relative interiors of the nonempty faces of A form a partition of A, that is a subdivision of A in mutually disjoint subsets.

To end, it should be noticed that a convex set A need not have p-dimensional faces for every p inferior or equal to the dimension of A. If, for instance, A is the closed ball $\overline{B(0,1)}$ of $\mathbb{R}^n$, all the points of the sphere $S_{n-1} = \{x \in \mathbb{R}^n \mid \|x\| = 1\}$ are extreme points of A. There is no nonempty face of A other than A itself and the points of S_{n-1}. If A is an affine set, A is the only nonempty face of A. The *p-simplices*, defined as convex hulls of $(p+1)$ affinely independent points of $\mathbb{R}^n$, are examples of convex sets which have q-dimensional faces for every $q \leq p$. Indeed, if $A = \mathrm{co}\{x^0, \ldots, x^p\}$ is such a p-simplex, the convex hull of any subset of $\{x^0, \ldots, x^p\}$ is an extreme subset of A.

3.1.2 Krein-Milman's theorem

Proposition 3.1.1. *A nonempty compact convex subset A of $\mathbb{R}^n$ has an extreme point.*

Proof. Consider the (continuous) real function defined on $\mathbb{R}^n$: $x \to \|x\|^2$. Let x^0 be a point which realizes the maximum on A of this function. We claim that x^0 is an extreme point of A. Indeed, if x^0 is not extreme, there exist in A two different points y and z such that $x^0 = (1/2)(y+z)$. From the definition of the Euclidean norm it follows that $\|x^0\|^2 < (1/2)(\|y\|^2 + \|z\|^2)$, which contradicts the inequalities $\|y\|^2 \leq \|x^0\|^2$ and $\|z\|^2 \leq \|x^0\|^2$. □

More generally,

Proposition 3.1.2. *A nonempty closed convex subset P of $\mathbb{R}^n$ has an extreme affine subset (that is, an extreme subset which is an affine subset of P). Each such subset of P is a maximal element for the inclusion relation in the set of all the affine subsets of P. Moreover, every extreme affine subset of P is parallel to the linear subspace $L := \{y \in \mathbb{R}^n \mid \{y\} + P = P\} = \mathrm{A}(P) \cap (-\mathrm{A}(P))$, where $\mathrm{A}(P)$ denotes the asymptotic cone of P (see Exercise 1.17 for the definition of $\mathrm{A}(P)$).*

Proof. The proof of the first assertion is by induction on the dimension p of P. If $p = 0$ or 1, the proposition is obviously true. Suppose it is true for $p \leq k$ and let $\dim P = k+1$. If $P = \mathrm{aff}\, P$, P is itself an extreme affine set. If $P \neq \mathrm{aff}\, P$, since P is closed, $P \setminus \mathrm{ri}(P) \neq \emptyset$. Thus let $x^0 \in P \setminus \mathrm{ri}(P)$. In view of Proposition 2.1.9, let H be a nontrivial supporting hyperplane to P at x^0. As $\dim(H \cap P) \leq k$ and since $H \cap P$ is closed and nonempty, by the induction hypothesis, $H \cap P$ contains an extreme affine subset M. Since $H \cap P$ is itself an extreme subset of P, we have that M is an extreme affine subset of P.
That an extreme affine set is a maximal element for the inclusion relation in the set of all the affine subsets of P is easily verified. Indeed, let F be an extreme affine set and G an affine subset of P which contains F. Given $x \in F$ and $y \in G$, $y \neq x$, let $z = x - \lambda(y-x)$ for some $\lambda > 0$. Since G is affine, $z \in G$; since F is an extreme affine subset, $y \in F$, which shows that G coincides with F.
That L, as defined in the proposition, is a linear subspace is also easily verified: $y \in L$ and $z \in L$ imply $y+z \in L$ and $-y \in L$; on the other hand, in view of the convexity of P, $y \in L$ implies $\lambda y \in L$, for every λ such

that $0 < \lambda \leq 1$. As $\{ky\} + P = P$ for every integer k, it follows that $y \in L$ implies $\lambda y \in L$, $\forall \lambda > 0$. Moreover, if $\mathrm{A}(P)$ denotes the asymptotic cone of P, it follows from the definitions of $\mathrm{A}(P)$ and L that $L = \mathrm{A}(P) \cap (-\mathrm{A}(P))$. We now show that if F is an extreme affine set, then $L = F - F$. Indeed, $F \subset F + L \subset P$ and F extreme affine imply $F = F + L$, thus $L \subset F - F$. Conversely, in view of the closedness of P, $F - F$ is contained in $\mathrm{A}(P)$; since $F - F$ is a linear subspace of $\mathbb{R}^n$, $F - F$ is contained in $L = \mathrm{A}(P) \cap (-\mathrm{A}(P))$. □

The linear space $L = \{y \in \mathbb{R}^n \mid \{y\} + P = P\}$ is called the *lineality space* of P. As shown in the previous proof, L is the largest linear subspace of $\mathbb{R}^n$ contained in the asymptotic cone $\mathrm{A}(P)$.

Proposition 3.1.3. *A nonempty, convex and compact subset of $\mathbb{R}^n$ is the closed convex hull of the set of its extreme points.*

Proof. Let $A \subset \mathbb{R}^n$ be nonempty, convex and compact and, in view of the previous proposition, let S be the nonempty set of the extreme points of A. Recall first that $\overline{\mathrm{co}}\, S$, the closed convex hull of S, is the smallest closed convex set containing S, equivalently the closure of the convex hull of S (see Corollary 1.2.1). Since A is convex and closed, one has $\overline{\mathrm{co}}\, S \subset A$. We now show by contraposition that $A \subset \overline{\mathrm{co}}\, S$. Indeed, if $x \in A$, $x \notin \overline{\mathrm{co}}\, S$, then it follows from Proposition 2.1.6 that there exists a non null linear functional f and a real number α such that $f(z) \leq \alpha < f(x)$, $\forall z \in \overline{\mathrm{co}}\, S$. Since A is compact, let $x^0 \in A$ be such that $f(y) \leq f(x^0)$ $\forall y \in A$. Then consider H, the hyperplane defined by the equation $f(y) = f(x^0)$. The set $H \cap A$ is nonempty, convex, compact and has, in view of Proposition 3.1.1, an extreme point that we denote by w. If H contains A, then $H \cap A = A$ and w is an extreme point of A. If H does not contain A, H is a nontrivial supporting hyperplane to A. In this case, it is easily verified that $H \cap A$ is a face of A, different from A. Thus, w, being an extreme point of $H \cap A$, is an extreme point of A. So in both cases, $w \in S \subset \overline{\mathrm{co}}\, S$ and $f(w) < f(x) \leq f(x^0) = f(w)$, which yields a contradiction. We have thus proved that $A \subset \overline{\mathrm{co}}\, S$, hence $A = \overline{\mathrm{co}}\, S$ □

The conclusion of Proposition 3.1.3 is the regular statement for a nonempty, convex and compact subset of a Hausdorff locally convex topological linear space. Notice that, without the finite dimension hypothesis, the very existence of extreme points would require Zorn's lemma in place of the geometric argument used in Proposition 3.1.1 or of the induction reasoning used in Proposition 3.1.2. Actually, the properties of $\mathbb{R}^n$ allow to obtain a sharper result due to Minkowski:

Proposition 3.1.4. *A nonempty, convex and compact subset A of $\mathbb{R}^n$ is the convex hull of the set of its extreme points.*

Proof. Let again S be the nonempty set of the extreme points of A. It follows from the previous proposition that $A = \overline{\mathrm{co}}\, S$. We will prove by induction on the dimension of A that $A = \mathrm{co}\, S$. It obviously holds if $\dim A = 0$, that is if $A = \{a\}$ for some $a \in \mathbb{R}^n$. Let us assume that the statement $B = \mathrm{co}\, T$ holds for any compact convex set B satisfying $\dim B < p$, if T is the set of extreme points of B. Assume now that $\dim A = p$. Let $x \in A$.

If $x \in \mathrm{ri}(A)$, we then have $x \in \mathrm{ri}(\overline{\mathrm{co}}\, S) = \mathrm{ri}(\overline{\mathrm{co}\, S}) = \mathrm{ri}(\mathrm{co}\, S)$ (see Corollary 1.2.4).

If x belongs to the relative boundary of A, in view of Proposition 2.1.9, let H be a nontrivial support hyperplane to A containing x and let f be a linear functional such that $f(z) \leq f(x)$, $\forall z \in A$, $f(z) < f(x)$ if $z \in \mathrm{ri}(A)$. We first remark that $H \cap A$ is a face of A and now prove that $\dim(H \cap A) < p$. Indeed, if $\{a^0, a^1, \ldots, a^q\}$ is a set of affinely independent points of $H \cap A$, $f(x^0) < f(x) = f(a^0) = f(a^1) = \ldots = f(a^q)$, and x^0 cannot be expressed as an affine combination of the points $a^0, a^1, \ldots, a^q$. This shows that $\dim A \geq (q+1)$, thus $\dim A > \dim(H \cap A)$. Applying now the induction hypothesis, $H \cap A$ is the convex hull of its extreme points. Since every extreme point of $H \cap A$ is an extreme point of A, we have proved that $x \in \mathrm{co}\, S$.

In both cases, $x \in \mathrm{co}\, S$. Hence $A \subset \mathrm{co}\, S \subset \overline{\mathrm{co}}\, S = A$, which shows that $A = \mathrm{co}\, S$. □

From now on, Proposition 3.1.4 will be referred to as Krein-Milman's theorem. Using the same ideas originating from Klee, it will be proved in Exercise 3.10 that more generally, every closed convex set of $\mathbb{R}^n$ containing no line is the convex hull of the union of its extreme points and its extreme rays.

3.2 Application to linear inequalities. Weyl's theorem

Let us first set two definitions:
A *polytope* is the convex hull of finitely many points. This definition includes *simplices* which are convex sets generated by finitely many affinely independent points.
A *polyhedral convex cone* is a pointed convex cone generated by finitely many points.

Proposition 3.2.1. *The following properties of a nonempty compact convex subset A of $\mathbb{R}^n$ are equivalent:*
(i) A is the intersection of some finite collection of closed half-spaces (that is, A is the set of solutions to finitely many linear inequations);
(ii) A is a polytope.

Proof. (i) implies (ii): Let us assume that $A = \{x \in \mathbb{R}^n \mid a^i \cdot x \leq b^i,\ i = 1, \ldots, m\}$, $a^i \in \mathbb{R}^n \setminus \{0\}$, $b^i \in \mathbb{R}$, $i = 1, \ldots, m$. In view of Krein-Milman's theorem, we have to prove that A has finitely many extreme points. For each extreme point x of A, define $N(x) = \{i \in \{1, \ldots, m\} \mid a^i \cdot x < b^i\}$. We claim that if x and y are two different extreme points of A, then $N(x) \neq N(y)$. Indeed if $N(x) = N(y) \neq \emptyset$, set

$$\theta = \min_{i \in N(x)} \frac{b^i - a^i \cdot x}{b^i - a^i \cdot y}.$$

By definition, $\theta > 0$ and $(b^i - a^i \cdot y)\theta \leq b^i - a^i \cdot x$, $\forall i = 1, \dots, m$. If $\theta \geq 1$ then for all $i = 1, \dots, m$, $(b^i - a^i \cdot y) \leq b^i - a^i \cdot x$, thus $a^i \cdot x \leq a^i \cdot y$. Since $a^i \cdot y < b^i \Leftrightarrow i \in N(x)$, there exists $\lambda > 1$ such that $a^i \cdot \big(\lambda x + (1-\lambda)y\big) < b^i$, $\forall i \in N(x)$, $a^i \cdot \big(\lambda x + (1-\lambda)y\big) = b^i$, $\forall i \notin N(x)$, hence $\lambda x + (1-\lambda)y \in A$, and $x = (1/\lambda)\big(\lambda x + (1-\lambda)y\big) + \big((\lambda - 1)/\lambda\big)y$ cannot be an extreme point of A. If $\theta < 1$, an easy calculation shows that $a^i \cdot \big(1/(1-\theta)\big)(x - \theta y) \leq b^i$, $\forall i = 1, \dots, m$, hence that $\big(1/(1-\theta)\big)(x - \theta y) \in A$. Then $x = \theta y + (1-\theta)\big(1/(1-\theta)\big)(x - \theta y)$ cannot be an extreme point of A.
If $N(x) = N(y) = \emptyset$, the affine set $\{z \in \mathbb{R}^n \mid a^i \cdot z = b^i,\ i = 1, \dots, m\}$, contained in A, contains x and y, thus has a dimension at least equal to 1, which contradicts the compactness of A.
We have then proved that each extreme point of A is associated with one and only one (possibly empty) subset of $\{1, \dots, m\}$. There are only finitely many subsets of the set $\{1, \dots, m\}$, thus there are only finitely many extreme points of A.

(ii) implies (i): Assume conversely that $A = \text{co}\{a^1, \dots, a^s\}$.
If $\dim A = n$, A has interior points (Proposition 1.2.3). Without loss of generality, one can assume that $0 \in \text{i}(A)$. In view of Proposition 2.2.3, the set $A^\circ = \{y \in \mathbb{R}^n \mid a^k \cdot y \leq 1,\ k = 1, \dots, s\}$ is compact. Applying the first part of the proof, let $y^1, \dots, y^r$ be such that $A^\circ = \text{co}\{y^1, \dots, y^r\}$. Then, since A is convex closed and contains 0, using Proposition 2.2.2, one sees that $A = A^{\circ\circ} = \{x \in \mathbb{R}^n \mid y^k \cdot x \leq 1,\ \forall k = 1, \dots, r\}$.
If $\dim A < n$, one can assume without loss of generality that $0 \in \text{ri}(A)$. Let W be the p-dimensional linear subspace generated by A and let $f : \mathbb{R}^p \to W$ be a linear isomorphism from $\mathbb{R}^p$ onto W. Setting $A' := f^{-1}(A)$, A' is a convex compact subset of $\mathbb{R}^p$, actually the convex hull of finitely many points $\{f^{-1}(a^1), \dots, f^{-1}(a^s)\}$. Applying in $\mathbb{R}^p$ the arguments given previously for the case where A has a full dimension, we get the existence of $d^1, \dots, d^r$ such that $A' = \{z \in \mathbb{R}^p \mid d^k \cdot z \leq 1,\ \forall k = 1, \dots, r\}$. Then $A = f(A') = \{x \in W \mid d^k \cdot f^{-1}(x) \leq 1,\ \forall k = 1, \dots, r\}$. Since $d^k \cdot f^{-1}(x) \leq 1$ is a linear inequation and W can be defined by linear equations, A is, as desired, the intersection of finitely many closed half-spaces. □

The extreme points of A are called *vertices* of A.

Proposition 3.2.2. *Let C be a pointed convex cone in $\mathbb{R}^n$. The following properties are equivalent:*
(i) C is the intersection of some finite collection of closed half-spaces determined by hyperplanes containing the origin (that is, C is the set of solutions to finitely many homogeneous linear inequations);
(ii) C is a polyhedral convex cone (that is a pointed convex cone generated by finitely many points).

Proof. (i) implies (ii): If C does not contain any line, one easily verifies the following assertions:
- $\dim C^\circ = n$;
- if $y \in \mathrm{i}(C^\circ)$, then $x \cdot y < 0,\ \forall x \in C,\ x \neq 0$;
- C is the pointed convex cone generated by the set $A_y := \{x \in C \mid x \cdot y = -1\}$;
- A_y is convex and compact.

Applying now to A_y the proposition 3.2.1, one has that A_y is the convex hull of finitely many points $\{x^1, \ldots, x^s\}$. C is obviously the pointed convex cone generated by the finite set $\{x^1, \ldots, x^s\}$.
If C contains lines, define the linear subspace $L := C \cap (-C)$. One easily verifies that $C = L + (C \cap L^\perp)$ where $L^\perp$ is the orthogonal complement of L. As previously proved, the pointed convex cone $C \cap L^\perp$ is generated by a finite set of points $\{c^1, \ldots, c^s\}$. Let then $f^1, \ldots, f^p$ be a basis of L. Obviously, C is the pointed convex cone generated by the set $\{c^1, \ldots, c^s, f^1, \ldots, f^p, -f^1, \ldots, -f^p\}$.

(ii) implies (i): Suppose now that C is generated by the finite set $\{c^1, \ldots, c^t\}$. Applying the first part of the proof to the closed convex cone $C^\circ = \{y \in \mathbb{R}^n \mid y \cdot c^k \leq 0,\ k = 1, \ldots, t\}$, one has that C° is generated by a finite system $\{y^1, \ldots, y^r\}$. As C is closed (Lemma 2.3.3, see also Exercise 2.18), $C = C^{\circ\circ} = \{x \in \mathbb{R}^n \mid x \cdot y^k \leq 0,\ \forall k = 1, \ldots, r\}$. □

The interest of Propositions 3.2.1 and 3.2.2 is to precise the geometric structure of the set of solutions to some finite system of linear inequalities:
- it is a polytope if this set is nonempty and compact;
- it is a polyhedral convex cone if the linear inequalities are homogeneous.

It remains to deal with the general case of finitely many non necessarily homogeneous linear inequalities when the nonempty set of solutions need not be compact.

Proposition 3.2.3. *Let P be the intersection of some finite collection of closed half-spaces (that is, the set of solutions to some finite system of linear inequalities). If P is nonempty, P can be expressed as the sum $P = A + C$, where A is a polytope and C is the polyhedral convex cone of the solutions to the system of the homogeneous inequalities associated with the system defining P. If P does not contain any line (that is, if $C \cap (-C) = \{0\}$), the vertices of A are precisely the extreme points of P.*

Proof. Suppose that $P = \{x \in \mathbb{R}^n \mid a^i \cdot x \leq b^i,\ i = 1, \ldots, m\}$. If P is compact or if the inequations are all homogeneous, in view of the two previous propositions, there is nothing to prove. Let us assume that for some i, $b^i \neq 0$ and that P is not compact. Defining in $\mathbb{R}^{n+1}$ the polyhedral convex cone $K = \{(x,t) \in \mathbb{R}^n \times \mathbb{R}_+ \mid a^i \cdot x - b^i t \leq 0,\ i = 1, \ldots, m\}$, one obviously has: $x \in P \Leftrightarrow (x,1) \in K$. In view of Proposition 3.2.2, K is finitely generated. Without loss of generality, we will assume in the whole proof that the set of generators is minimal and equal to the union of two sets $\{(c^1,1), \ldots, (c^q,1)\}$ and $\{(c^{q+1},0), \ldots, (c^r,0)\}$ with $c^1, \ldots, c^r$ in $\mathbb{R}^n$, so that

$x \in P \Leftrightarrow x = \sum_{k=1}^{r} \lambda_k c^k$ for some system of $\lambda_k \geq 0,\ k = 1, \ldots, r$, satisfying $\sum_{k=1}^{q} \lambda_k = 1$. Note that necessarily both sets $\{1, \ldots, q\}$ and $\{q+1, \ldots, r\}$ are nonempty, the first one because P is nonempty and the second one because P is not compact. Then let A be the polytope generated by $\{c^1, \ldots, c^q\}$ and let C be the polyhedral pointed convex cone generated by $\{c^{q+1}, \ldots, c^r\}$. Obviously $P = A + C$.
Moreover, for all $k = q+1, \ldots, r$, $(c^k, 0) \in K \Rightarrow a^i \cdot c^k \leq 0,\ i = 1, \ldots, m$. Conversely, if $a^i \cdot x \leq 0,\ i = 1, \ldots, m$, then $(x, 0) \in K$. From $(x,0) = \sum_{k=1}^{q} \lambda_k (c^k, 1) + \sum_{k=q+1}^{r} \lambda_k (c^k, 0)$, for some $\lambda_k \geq 0,\ k = 1, \ldots, r$, one deduces that $x = \sum_{k=q+1}^{r} \lambda_k c^k$, which shows that C is the set of solutions to the system of homogeneous linear inequalities $a^i \cdot x \leq 0,\ i = 1, \ldots, m$. It is worth noticing that C is precisely the asymptotic cone of P (for a definition of the asymptotic cone (recession cone) of a convex subset of $\mathbb{R}^n$, see Exercise 1.18).
Suppose now that P does not contain any line. Let us first notice that $c^1, \ldots, c^q$ are the extreme points of A (the system of generators of K was supposed to be minimal). Let us show that $c^1, \ldots, c^q$ are extreme points of P. Indeed, if for some $k \in \{1, \ldots, q\}$, c^k can be expressed as $c^k = \lambda(a + c) + (1 - \lambda)(b + d)$ with $a, b \in A$, $c, d \in C$, $0 < \lambda < 1$, setting $a = \sum_{h=1}^{q} \mu_h c^h$, $b = \sum_{h=1}^{q} \nu_h c^h$, with $\mu_h \geq 0$, $\nu_h \geq 0$ and $\sum_{h=1}^{q} \mu_h = 1$, $\sum_{h=1}^{q} \nu_h = 1$, we then have

$$c^k = \sum_{h=1}^{q} (\lambda \mu_h + (1 - \lambda)\nu_h) c^h + \lambda c + (1 - \lambda) d.$$

Let us distinguish two cases:

if $\lambda \mu_k + (1 - \lambda)\nu_k < 1$, it is easily verified that $(c^k, 1)$ can be dropped from the set of generators of K, which contradicts the assumption that the system of generators of K is minimal;

if $\lambda \mu_k + (1 - \lambda)\nu_k = 1$, then $\mu_k = \nu_k = 1$, which implies $c^k = a = b$ and $\lambda c + (1 - \lambda) d = 0$. In view of the hypothesis $C \cap (-C) = \{0\}$, we have then $c = d = 0$ and $a + c = b + d$, which shows that c^k is an extreme point P.
On the other hand, recalling that $P = A + C$, it is easily seen that every extreme point of P is an extreme point of A. The proof is now complete. □

Corollary 3.2.1. *The following properties of a nonempty convex subset P of $\mathbb{R}^n$ are equivalent:*
(i) P is the set of solutions to some finite system of linear inequalities;
(ii) P can be expressed as the sum $P = A + C$, where A is a polytope and C is the polyhedral convex cone of the solutions to the system of the homogeneous inequalities associated with the system defining P.

Proof. (i) implies (ii) was proved in Proposition 3.2.3.
(ii) implies (i): Recall that A is compact, C is closed (Lemma 2.3.3 or Proposition 3.2.2), P is closed as the sum of a closed set and of a compact set

(Exercise 1.5). On the other hand, it is easily verified that $P^\circ = A^\circ \cap C^\circ$. Indeed, the inclusions $(A^\circ \cap C^\circ) \subset P^\circ$ and $P^\circ \subset A^\circ$ are obvious. The inclusion $P^\circ \subset C^\circ$ follows from the fact that C is a cone. Since A and C have finite numbers of generators, P° is the intersection of some finite collection of closed half-spaces. In view of Proposition 3.2.3, P° can be expressed as the sum $P^\circ = B + D$, where B is a polytope and D is a polyhedral convex cone. Then $P = P^{\circ\circ} = (P^\circ)^\circ = (B + D)^\circ = B^\circ \cap D^\circ$. Since B and D are finitely generated, P is obviously the set of solutions to some finite system of linear inequalities. □

Let us call *consistent* a finite system of linear inequalities which has a nonempty set of solutions. A *polyhedral convex set* is by definition the set of solutions to some consistent finite system of linear inequalities or, equivalently, a set which can be expressed as the sum of a polytope and of a polyhedral convex cone. Using one or the other characterization for polyhedral convex sets, it is easily verified that if P, P_1, P_2 are polyhedral convex sets of $\mathbb{R}^n$, and if $\lambda \in \mathbb{R}$, then $P_1 \cap P_2$, $P_1 + P_2$, λP, $\mathrm{A}(P)$, $\mathrm{K}(P) \cup \mathrm{A}(P)$ (where $\mathrm{K}(P) = \bigcup\{\lambda P \mid \lambda > 0\}$ and $\mathrm{A}(P)$ denote respectively the (convex) cone generated by P and the asymptotic cone of P) are polyhedral convex sets.

As a consequence, the following proposition refines the second separation theorem:

Proposition 3.2.4. *If P_1 and P_2 are nonempty disjoint polyhedral convex sets, there exists a hyperplane separating P_1 and P_2 strictly.*

Proof. We have $0 \notin (P_2 - P_1)$ and $P_2 - P_1$, being a polyhedral convex set, is closed. It is then possible to apply Proposition 2.1.8 and to find $p \in \mathbb{R}^n \setminus \{0\}$ and $\delta \in \mathbb{R}$ such that $p \cdot (y - x) > \delta > 0$, for all $x \in P_1$, $y \in P_2$. Let $\alpha = \sup\{p \cdot x \mid x \in P_1\}$, $\beta = \inf\{p \cdot y \mid y \in P_2\}$ and γ be such that $\alpha < \gamma < \beta$. The hyperplane of equation $p \cdot z = \gamma$ separates strictly P_1 and P_2. □

Once again, recalling that polyhedral convex sets are closed, we have also the following immediate result:

Proposition 3.2.5. *Let P be a nonempty polyhedral convex set. Let $K(P)$ and $\mathrm{A}(P)$ denote respectively the (convex) cone generated by P and the asymptotic cone of P. Then $\overline{\mathrm{K}(P)} = K(P) \cup \mathrm{A}(P)$. In particular, if $0 \in P$, then $\mathrm{K}(P)$ is closed and polyhedral.*

Proof. Since P is closed, it follows from the definition of $\mathrm{A}(P)$ (see Exercise 1.17) that $\mathrm{A}(P) \subset \overline{\mathrm{K}(P)}$. We then have $\mathrm{K}(P) \subset (\mathrm{K}(P) \cup \mathrm{A}(P)) \subset \overline{\mathrm{K}(P)}$. Since $\mathrm{K}(P) \cup \mathrm{A}(P)$ is a polyhedral convex set, hence a closed convex set, $\overline{\mathrm{K}(P)} = \mathrm{K}(P) \cup \mathrm{A}(P)$. If $0 \in P$, $\mathrm{A}(P) \subset P \subset \mathrm{K}(P)$ (see also Exercise 1.17) and $\overline{\mathrm{K}(P)} = \mathrm{K}(P) \cup \mathrm{A}(P) = \mathrm{K}(P)$. □

The following separation result and its rather intricate proof are borrowed from Rockafellar [51]. This result will be used in Chapter 7.

Proposition 3.2.6. *Let P and C be nonempty convex sets such that P is polyhedral and $P \cap \mathrm{ri}(C) = \emptyset$. Then there exists a hyperplane separating P and C which does not contain C.*

Proof. Let $F = \mathrm{aff}\, C$ be the affine hull of C. Set $D := P \cap F$. We first notice that the assumption $P \cap \mathrm{ri}(C) = \emptyset$ is equivalent to $D \cap \mathrm{ri}(C) = \emptyset$. If $D = \emptyset$, then P and F, being polyhedral convex sets, can be strictly separated (Proposition 3.2.4), which implies the conclusion of the proposition. Let us now assume that $D \neq \emptyset$. In view of $D \cap \mathrm{ri}(C) = \emptyset$, let H be a hyperplane which properly separates D and C (Proposition 2.1.10). Note that H does not contain C: indeed, if not, H contains also D, which contradicts the fact that H separates properly D and C. Define now C' as the intersection of F with the closed half-space containing C. If $C' \cap P = \emptyset$, then then P and C', being polyhedral convex sets, can be strictly separated (Proposition 3.2.4), which implies that C and P can be strictly, hence properly, separated. Assume then that $C' \cap P \neq \emptyset$, thus $P \cap F \cap H \neq \emptyset$. Translating all the sets if necessary, one can assume that $0 \in P \cap F \cap H$. Define $\mathrm{K}(P)$ as the convex cone generated by P. As noticed in Proposition 3.2.5, $\mathrm{K}(P)$ is a polyhedral convex cone. Moreover, $\mathrm{K}(P) \cap \mathrm{ri}(C') = \emptyset$. Let $P' = \mathrm{K}(P) + (H \cap F)$. Then P' is a polyhedral convex cone containing P and such that $P' \cap C' = H \cap F$. Expressing P' as a finite intersection of closed half-spaces determined by hyperplanes containing the origin, it is then easily seen that at least one of these closed half-spaces does not meet $\mathrm{ri}(C')$. The boundary hyperplane of this half-space separates P' and and $\mathrm{ri}(C')$ and does not contain C', thus separates P and C and does not contain C. □

3.3 Extreme points and extremal subsets of a polyhedral convex set

In this section, we indicate how to compute the decomposition of a polyhedral convex set as $P = A + C$ when P is given as the set of solutions to some finite system of linear inequalities.

Let us first recall a definition: Given a system of linear equations $Bx = b$ for some $(m \times n)$-matrix B and some $b \in \mathbb{R}^m$, one says that the system has a *rank* equal to r if the matrices B and $(\,B \quad b\,)$ have a same rank equal to r, that is, if $\mathrm{rank}(B) = \mathrm{rank}\,(\,B \quad b\,) = r$.

Proposition 3.3.1. *Let $P = \{x \in \mathbb{R}^n \mid a^i \cdot x \leq b^i,\ i = 1, \ldots, m\}$ be a polyhedral convex set. If the system $a^i \cdot x = b^i,\ i = 1, \ldots, m$ has a subsystem of rank n, let x be the unique solution to this subsystem. Then either x is an extreme point of P or x does not belong to P. Conversely, every extreme point of P can be expressed as the unique solution to such a subsystem.*

Proof. Let $\{i_1, \ldots, i_n\}$ be such that the subsystem $a^{i_j} \cdot x = b^{i_j},\ j = 1, \ldots n$ has a rank equal to n and let x be the (necessarily unique) solution to this

subsystem (note that necessarily $m \geq n$).
If $x \in P$ and if $x = \lambda y + (1 - \lambda)z$ for some λ, $0 < \lambda < 1$ and some $y, z \in P$, then from $a^{i_j} \cdot y \leq b^{i_j}$, $a^{i_j} \cdot z \leq b^{i_j}$, $a^{i_j} \cdot x = b^{i_j}$, $\forall j = 1, \ldots, n$, it follows that $a^{i_j} \cdot y = a^{i_j} \cdot z = b^{i_j}$, $\forall j = 1, \ldots n$, which in turn proves that $x = y = z$, thus that x is an extreme point of P.
Assume conversely that x is an extreme point of P and define $I = \{i \in \{1, \ldots, m\} \mid a^i \cdot x = b^i\}$, $J = \{i \in \{1, \ldots, m\} \mid a^i \cdot x < b^i\}$. We will prove by contraposition that x is the unique solution to the system $a^i \cdot z = b^i$, $i \in I$. Indeed, let $y \neq x$ be another solution to $a^i \cdot z = b^i$, $i \in I$. If $J \neq \emptyset$, let $\varepsilon > 0$ be such that $\|z - x\| < \varepsilon \Rightarrow a^i \cdot z < b^i$, $\forall i \in J$ and let λ be such that $0 < |\lambda| < (\varepsilon/\|y - x\|)$. If $J = \emptyset$, take any $\lambda \neq 0$. In both cases, $x + \lambda(y - x) \in P$, $x - \lambda(y - x) \in P$ and $x = (1/2)\big(x + \lambda(y - x)\big) + (1/2)\big(x - \lambda(y - x)\big)$, which contradicts the assumption that x is an extreme point of P. Thus x is the unique solution to the system $a^i \cdot z = b^i$, $i \in I$ which has necessarily a rank equal to n. □

Proposition 3.3.2. *Let $P = \{x \in \mathbb{R}^n \mid a^i \cdot x \leq b^i,\ i = 1, \ldots, m\}$ be a polyhedral convex set. If the system $a^i \cdot x = b^i$, $i = 1, \ldots m$ has a subsystem of rank $n - 1$, let Δ be the line of the solutions to this subsystem. Then either $P \cap \Delta$ has at most one point, or $P \cap \Delta$ is a 1-dimensional face of P. Conversely, every line which contains a 1-dimensional face of P is the solution to such a subsystem.*

Proof. Let $\{i_1, \ldots, i_{n-1}\}$ be such that the subsystem $a^{i_j} \cdot z = b^{i_j}$, $j = 1, \ldots, (n - 1)$ has a rank equal to $n - 1$ and let Δ be the line of the solutions to this subsystem. Let us show that if $P \cap \Delta$ has at least two points, then $P \cap \Delta$ is a (1-dimensional) face of P. Indeed, let $x \in]u, v[$, $u, v \in P$, $x \in P \cap \Delta$. Obviously, $a^{i_j} \cdot u = b^{i_j}$ and $a^{i_j} \cdot v = b^{i_j}$, $\forall j = 1, \ldots, (n - 1)$, thus $u, v \in P \cap \Delta$, which shows that $P \cap \Delta$ is a face of P. This face which has at least two points is 1-dimensional.
Conversely, assume that Δ is a line and that $P \cap \Delta$ is a 1-dimensional face of P. Let $x \in \mathrm{ri}(P \cap \Delta)$, and let $I_x = \{i \in \{1, \ldots, m\} \mid a^i \cdot x = b^i\}$, $J_x = \{i \in \{1, \ldots, m\} \mid a^i \cdot x < b^i\}$. Since $x \in \mathrm{ri}(P \cap \Delta)$, there exist u and v in $P \cap \Delta$ such that $x \in]u, v[$ and Δ is the line going through u and v, which shows that $\Delta \subset \{z \in \mathbb{R}^n \mid a^i \cdot z = b^i,\ \forall i \in I_x\}$. On the other hand, if $y \neq x$ is a solution of the system $a^i \cdot z = b^i$, $\forall i \in I_x$, one shows as in the proof of the previous proposition that there exists $\lambda \neq 0$ such that $x + \lambda(y - x) \in P$, $x - \lambda(y - x) \in P$ and $x = (1/2)\big(x + \lambda(y - x)\big) + (1/2)\big(x - \lambda(y - x)\big)$. Since $P \cap \Delta$ is a face of P, $x + \lambda(y - x) \in \Delta$, $x - \lambda(y - x) \in \Delta$, hence $y \in \Delta$ and $\Delta = \{z \in \mathbb{R}^n \mid a^i \cdot z = b^i,\ \forall i \in I_x\}$. The system $a^i \cdot z = b^i$, $\forall i \in I_x$ has necessarily a rank equal to $n - 1$. □

It is worth noticing that if $P \cap \Delta$ is a 1-dimensional face of P which is not the whole line Δ, then $P \cap \Delta$ is either a closed segment whose endpoints are extreme points of P or an extreme ray issued from an extreme point of P.

Exercises

Exercise 3.1 (Faces of a convex set). Let A be a convex subset of $\mathbb{R}^n$. Prove the following properties of extreme subsets of A:
a) If x is an extreme point of A then $A \setminus \{x\}$ is convex. More generally, if F is a face of A then $A \setminus F$ is convex.
b) Every intersection of faces of A is a face of A. Every face of a face of A is a face of A.
c) If D is a convex subset of A and if F is a face of A such that $\operatorname{ri}(D) \cap F \neq \emptyset$, then $D \subset F$.
d) If F is a face of A, one has: $F = A \cap \overline{F}$
(Hint: Apply c) to the set $D = A \cap \overline{F}$).
In particular, if A is closed then F is closed.
e) Every face of A distinct from A is contained in the relative boundary of A
(Hint: Apply c) to the set $D = \operatorname{ri}(A)$).
Then $\dim F < \dim A$.
f) Two distinct faces of A have disjoint relative interiors.
g) For every point x of the relative boundary of A, if H is a nontrivial supporting hyperplane of $\overline{A}$ containing x, then $H \cap A$ is a face of A.
h) For every $x \in A$, if $F_A(x)$ is the intersection of the faces of A containing x, then $x \in \operatorname{ri}(F_A(x))$.
i) $F_A(x) = \{z \in A \mid [x - \lambda(z - x), x + \lambda(z - x)] \subset A, \text{ for some } \lambda > 0\}$
(Hint: This set is convex. Use c) and h)).
j) If $V_A(x) = \operatorname{aff} F_A(x)$, then

$$V_A(x) = \{z \in \mathbb{R}^n \mid [x - \lambda(z - x), x + \lambda(z - x)] \subset A, \text{ for some } \lambda > 0\}.$$

Hence, $F_A(x) = A \cap V_A(x)$. In particular, x is an extreme point of A if and only if $A \cap V_A(x) = \{x\}$.
k) The collection of the relative interiors of the faces of A is a partition of A in mutually disjoint subsets of A
(Hint: Use f) and h)).

Exercise 3.2. Let $f : \mathbb{R}^n \to \mathbb{R}^m$ be an affine map and let C be a convex subset of $\mathbb{R}^n$. If $x \in C$ and if $y = f(x)$, using the notations of Exercise 3.1, show that $f\big(V_C(x)\big) \subset V_{f(C)}(y)$. If x is an extreme point of $C \cap f^{-1}(y)$, show that $\dim V_C(x) \leq \dim V_{f(C)}(y)$
(Hint: Show first that if x is an extreme point of $C \cap f^{-1}(y)$, then $u \in V_C(x)$ and $f(u) = f(x)$ imply $u = v$).

Exercise 3.3 (Shapley-Folkman theorem). The goal of this exercise is to show that the (vector) sum of "many" subsets of $\mathbb{R}^n$ "tends" to be convex. The ideas of the proof are borrowed from Ichiishi (1983).
Let $(S^i)_{i \in I}$ be a finite collection of compact subsets of $\mathbb{R}^n$. Define $C = \prod_{i \in I} \operatorname{co} S^i$ and $f : (\mathbb{R}^n)^I \to \mathbb{R}^n$ by $f(x) = \sum_{i \in I} x_i$. Show that every y of $f(C)$ has a representation $y = \sum_{i \in I} x_i$ such that $x_i \in \operatorname{co} S^i$ for all $i \in I$ and

such that $x_i \notin S^i$ for at most n of the indices i.
(Hint: Using the previous exercise, show that for some $x = (x_i)_{i\in I} \in C \cap f^{-1}(y)$, $\dim V_C(x) \leq n$, hence that x_i is an extreme point of $\operatorname{co} S^i$, thus a point of S^i, for all except at most n of the indices i.)
Prove that the result still holds true if the S^i are not assumed to be compact.

Exercise 3.4. Prove that a point x is an extreme point of the convex set A if and only if the relations $x = (1/2)y + (1/2)z$ together with $y \in A$, $z \in A$ imply $x = y = z$.

Exercise 3.5. Let $C = \{(x_1, x_2, x_3) \in \mathbb{R}^3 \mid x_3 = 0 \text{ and } x_1^2 + x_2^2 - 2x_1 = 0\}$, $B = C \cup \{(0,0,1)\} \cup \{(0,0,-1)\}$ and let $A = \operatorname{co} B$ be the convex hull of B. Show that A is compact. What are the extreme points of of A. Deduce that in $\mathbb{R}^3$, the set of extreme points of a compact convex set need not be closed. What in $\mathbb{R}^2$?

Exercise 3.6. Let Ω_2 be the set of (2×2)-matrices $A = (a_{ij})$ satisfying $a_{ij} \geq 0$, $i,j = 1,2$, $\sum_{i=1}^2 a_{ij} = 1$, $j = 1,2$, $\sum_{j=1}^2 a_{ij} = 1$, $i = 1,2$. Show that:
a) Ω_2 is a convex subset of the linear space of all (2×2)-matrices;
b) $I = \begin{pmatrix} 1 & 0 \\ 0 & 1 \end{pmatrix}$ and $P = \begin{pmatrix} 0 & 1 \\ 1 & 0 \end{pmatrix}$ are extreme points of Ω_2;
c) any member of Ω_2 is a convex combination of I and P which are the only extreme points of Ω_2.
The set Ω_2 is called the set of the 2-dimensional bi-stochastic matrices. The set Ω_n can be analogously defined. A $(n \times n)$-matrix $P = (p_{ij})_{\substack{i=1,\dots,n \\ j=1,\dots,n}}$ is said to be a *permutation matrix* if there exists a permutation σ of $\{1, \dots, n\}$ such that $p_{ij} = 1$ if $j = \sigma(i)$ and 0 otherwise. Prove more generally that the set of the extreme points of Ω_n is the set of permutation matrices.

Exercise 3.7. For $\lambda > 0$, let $B = \{x \in \mathbb{R}^n \mid |x_k| \leq \lambda, \ \forall k = 1, \dots, n\}$. Show that the extreme points of B can be expressed as $z = \sum_{k=1}^n \lambda_k \lambda e^k$, where for every $k = 1, \dots, n$, $\lambda_k \in \{-1, +1\}$ and e^k is the kth vector of the canonical basis of $\mathbb{R}^n$. Identify the polar set B° of B.

Exercise 3.8. Let A be a closed convex subset of $\mathbb{R}^n$. Show that if A has an extreme point then A is line-free.
(Hint: If A contains a line L, one can assume without loss of generality that $0 \in L \subset A$. For $y \in L$ and $x \in A \setminus L$, show that for every $\lambda > 0$, $(1 - (1/(\lambda+1)))x + (1/(\lambda+1))\lambda y) \in A$, thus that $x + y \in A$.)
Prove then that a nonempty closed convex subset of $\mathbb{R}^n$ has an extreme point if and only is it is line-free.

Exercise 3.9. Consider in $\mathbb{R}^2$ the set P of the solutions to the system:

$$\begin{aligned} 2x_1 - x_2 &\leq 4 \\ x_1 + 2x_2 &\leq 3 \\ x_1 \geq 0, \quad x_2 &\geq 0. \end{aligned}$$

Compute the extremal points and the extremal directions of P.

Exercise 3.10. Same questions for P in $\mathbb{R}^2$ defined by:

$$\begin{array}{rcrcl} 2x_1 & - & x_2 & \geq & -2 \\ x_1 & - & 2x_2 & \leq & 3 \\ x_1 & + & 2x_2 & \geq & 11 \\ x_1 \geq 0, & & x_2 & \geq & 0. \end{array}$$

Exercise 3.11. Consider in $\mathbb{R}^n$ a closed convex cone, C, different from $\{0\}$ and which does not contains any line. The goal of this exercise is to show that C is the convex hull of its extreme rays.
a) Let C° be the polar cone of C. Show that $\dim C^\circ = n$ thus that C° has a nonempty interior.
b) Given $u \in \mathrm{i}(C^\circ)$, define $\Delta = \{x \in C \mid x \cdot u = -1\}$.
If $z \neq 0$ is in C, show that $z \cdot u < 0$. Then show that Δ is nonempty, convex and compact and that $C = \bigcup_{\lambda \geq 0} \lambda \Delta$.
c) Let S be the set of the extreme points of Δ. If $x \in S$, show that the half-line issued from 0 and passing through x is an extreme ray of C and that every extreme ray of C can be obtained in this way.
d) Conclude that C is the convex hull of its extreme rays.

Exercise 3.12 (Klee's Theorem). Using an induction argument on the dimension of the set, show that every nonempty closed convex subset B of $\mathbb{R}^n$ which does not contain any line is the convex hull of its extreme points and extreme rays.
Noticing that every extreme ray of B is issued from an extreme point of B and that, in view of the closedness of B, every extreme direction of B belongs to the asymptotic cone $\mathrm{A}(B)$ (for the definition and the main properties, see Exercise 1.17), conclude that $B = \mathrm{co}(\mathrm{ext}(B)) + \mathrm{A}(B)$, where $\mathrm{ext}(B)$ denotes the set of all extreme points of B.

Exercise 3.13. Let P be a closed convex subset of $\mathbb{R}^n$ and let

$$L_P = \{x \in \mathbb{R}^n \mid \{x\} + P = P\}$$

be the *lineality space* of P. Denote by $L_P^\perp$ the orthogonal complement to L_P.
a) Show that $P = L_P + B_P$ where $B_P = P \cap L_P^\perp$. Show that B_P is a closed convex subset of $\mathbb{R}^n$ which does not contain any line.
b) Using Exercise 3.12, prove the following decomposition formula for P:

$$P = L_P + \mathrm{A}(B_P) + \mathrm{co}(\mathrm{ext}\, B_P).$$

This formula is for a closed convex subset of $\mathbb{R}^n$ an abstract equivalent of the decomposition obtained in Proposition 3.2.3 for a polyhedral convex set. It is referred to by Holmes (1975) as the **Klee-Minkowski-Hirsch-Hoffman-Goldman-Tucker theorem**.

Exercise 3.14. For a given convex subset A of $\mathbb{R}^n$, $x \in A$ is said to be an *exposed point* (a point of *strict convexity*) of A if A has a support hyperplane

H such that $H \cap A = \{a\}$.
a) Show that an exposed point x of A is an extreme point of A.
b) Give an example in $\mathbb{R}^2$ showing that an extreme point of a convex set need not be exposed.
Let us now assume that A is convex and compact.
c) For any point $p \in \mathbb{R}^n$, define $A_p = \{x \in A \mid \|y - p\| \leq \|x - p\|,\ \forall y \in A\}$. Show that every point x of A_p is an exposed point of A_p.
d) Setting $P = \bigcup\{A_p \mid p \in \mathbb{R}^n\}$, prove then that $A = \overline{\text{co}}\, P$.
(Hint: If $z \in A \setminus \overline{\text{co}}\, P$, define x to be the projection of z onto $\overline{\text{co}}\, P$ and $y = (1/2)(x + z)$. Consider the increasing sequence of open balls B^n with center $u^n = y + n(x - y)$ and radius $\|u^n - y\|$. Using Exercise 2.3, prove that $\overline{\text{co}}\, P \subset \bigcap_{n \in \mathbb{N}} B^n$. Using the compactness of $\overline{\text{co}}\, P$, prove then that for some $n_0 \in \mathbb{N}$, $\overline{\text{co}}\, P \subset B^{n_0}$ and obtain a contradiction with the definition of P.)
e) Deduce from the previous result that every extreme point of A is the limit of a sequence of exposed points of A. In other words, denoting by $\exp A$ the set of exposed points of A, $\text{ext}\, A \subset \overline{\exp A}$.

4. Linear Programming

A *linear programming* problem consists in finding the maximum (resp. minimum) value of a linear functional, subject to a finite number of linear constraints. If c and a^i, $i = 1, \ldots, m$ are elements of $\mathbb{R}^n$, and if $b = (b_1, \ldots, b_m)$ belongs to $\mathbb{R}^m$, the most general form of a linear programming problem is the following:

Maximize (resp. minimize)

$$f(x) = c \cdot x$$

subject to the conditions

$$a^i \cdot x \begin{pmatrix} \leq \\ = \\ \geq \end{pmatrix} b_i, \quad i = 1, \ldots, m$$

$$x \in \mathbb{R}^n.$$

The linear functional $f(x)$ is called the *objective function.* The linear equations and inequations are called *constraints* of the problem. The set of points that satisfy these linear equations or inequations is called *feasible set.* An element of this set is a *feasible solution*; it is an *optimal solution* if it solves the maximization (resp. minimization) problem. Let (P) and (P') be two optimization problems, we recall that (P) and (P') are *equivalent* if their sets of solutions are equal.

This chapter gives a brief introduction to linear programming, putting the emphasis on optimality conditions and the duality theory. For this, we study different forms (*canonical*, *standard*) of the problem given above in its general form, making certain conventions on the direction of the inequalities and / or introducing additional variables in order to fit with positivity constraints. The subject of algorithms for solving the problem is large and fast growing. For didactical reasons, we limit ourselves here to the simplex method. Minkowski-Farkas' lemma (Chapter 2) is the main tool as well for establishing necessary and sufficient conditions of optimality as for the duality theory in linear programming. Together with their using, the simplex method for computing optimal solutions relies on the geometry of polyhedral convex sets (Chapter 3).

To end, let us notice that linear programming problems are particular cases of convex programming problems which will be studied in a subsequent chapter. They are studied here independently. As it will be seen later, applying to linear programming results of the duality theory for convex programming would furnish weaker results than the ones given in this chapter.

4.1 Necessary and sufficient conditions of optimality

4.1.1 Necessary and sufficient conditions of optimality

Let f^k (resp. f'^k) $k = 1,\dots,q$, h^k (resp. h'^k) $k = 1,\dots,r$, g (resp. g') be linear functional on $\mathbb{R}^n$, α_k (resp. α'_k) $k = 1,\dots,q$, β_k (resp. β'_k) $k = 1,\dots,r$ be real numbers. We now consider linear programming problems expressed in the following form:

$$(I) \qquad \begin{array}{l} \text{Minimize } g(x) \text{ subject to} \\ f^k(x) = \alpha_k, \ k = 1,\dots,q \\ h^k(x) \geq \beta_k, \ k = 1,\dots,r \\ x \in \mathbb{R}^n. \end{array}$$

or in the following equivalent form:

$$(I') \qquad \begin{array}{l} \text{Maximize } g'(x) \text{ subject to} \\ f'^k(x) = \alpha'_k, \ k = 1,\dots,q \\ h'^k(x) \leq \beta'_k, \ k = 1,\dots,r \\ x \in \mathbb{R}^n. \end{array}$$

Proposition 4.1.1. *For a feasible solution $\overline{x}$ to be an optimal solution of Problem (I) (resp. (I')), it is both necessary and sufficient that there exist $\mu_1,\dots,\mu_q \in \mathbb{R}$ and $\nu_1,\dots,\nu_r \in \mathbb{R}_+$ such that g (resp. g') $= \sum_{k=1}^q \mu_k f^k + \sum_{k=1}^r \nu_k h^k$, with $\nu_k = 0$ if $h^k(\overline{x}) > \beta_k$ (resp. $h'^k(\overline{x}) < \beta'_k$).*

Proof. Let Δ be the feasible set of Problem (I) and let $\overline{x} \in \Delta$. To $\overline{x}$ we associate the set $I(\overline{x}) = \{k \in \{1,\dots,r\} \mid h^k(\overline{x}) = \beta_k\}$. It is worth noticing that the condition stated in the lemma can be rewritten:

$$\exists \mu_1,\dots,\mu_q \in \mathbb{R},\ \exists \nu_1,\dots,\nu_r \in \mathbb{R}_+ : g = \sum_{k=1}^q \mu_k f^k + \sum_{k \in I(\overline{x})} \nu_k h^k \tag{4.1}$$

Let us first assume that $\overline{x}$ is an optimal solution. Then we claim that

$$\left.\begin{array}{l} f^k(y) = 0,\ \forall k = 1,\dots,q \\ h^k(y) \geq 0,\ \forall k \in I(\overline{x}) \end{array}\right\} \Rightarrow g(y) \geq 0.$$

Indeed if $y \neq 0$ is a solution of the system at the left, then $x+\lambda y \in \Delta$ for some $\lambda > 0$ small enough. In view of the optimality of $\overline{x}$, $g(\overline{x}+\lambda y) = g(\overline{x})+\lambda g(y) \geq$

$g(\overline{x})$, which implies, as previously claimed, $g(y) \geq 0$. Applying now Corollary 2.3.3 to Proposition 2.3.1, we get that Condition 4.1 is necessary for the optimality of $\overline{x}$.
Conversely, assume Condition 4.1. Let $x \in \Delta$ and define $y \in \mathbb{R}^n$ such that $x = \overline{x} + y$. Then $f^k(y) = f^k(x) - f^k(\overline{x}) = \alpha_k - \alpha_k = 0,\ \forall k = 1, \ldots, q$ and $h^k(y) = h^k(x) - h^k(\overline{x}) \geq \beta_k - \beta_k = 0,\ \forall k \in I(\overline{x})$. Since $g = \sum_{k=1}^{p} \mu_k f^k + \sum_{k \in I(\overline{x})} \nu_k h^k$, we get $g(y) \geq 0$, that is: $g(x) \geq g(\overline{x})$, which shows that $\overline{x}$ is an optimal solution to Problem (I).
The proof is obviously the same for Problem (I'). □

4.1.2 Application to solving linear programming problems

Let us now consider a linear programming problem of type (I):

$$(I) \qquad \begin{array}{l} \text{Minimize } g(x) \text{ subject to} \\ f^k(x) = \alpha_k,\ k = 1, \ldots, q \\ h^k(x) \geq \beta_k,\ k = 1, \ldots, r \\ x \in \mathbb{R}^n. \end{array}$$

If the feasible set P is nonempty, it follows from Proposition 3.2.3 that the set P is the geometric sum of a polytope, A, and of a polyhedral convex cone, C, that is: $P = A + C$. Let $\{a^1, \ldots, a^r\}$ be the set of extreme points of A and $\{c^1, \ldots, c^s\}$ be a set of generators of C.

Proposition 4.1.2. *If the objective function is bounded below on $P \neq \emptyset$, then there are optimal solutions to Problem (I), at least one of which is an extreme point of A (an extreme point of P if P does not contains any line). Moreover, every optimal solution in A belongs to the convex hull of the extreme points of A that are optimal solutions to (I).*

Proof. Assume that g is bounded below on $P \neq \emptyset$. Every point x of P can be expressed as $x = \sum_{k=1}^{r} \alpha_k a^k + \sum_{k=1}^{s} \gamma_k c^k$, with $\alpha_k \geq 0,\ k = 1, \ldots, r$, $\sum_{k=1}^{r} \alpha_k = 1,\ \gamma_k \geq 0,\ k = 1, \ldots, s$, so that $g(c^k) \geq 0,\ \forall k = 1, \ldots, s$ and $g(x) \geq g(\sum_{k=1}^{r} \alpha_k a^k)$. Let x^0 be such that the continuous function g has a minimum at x^0 on the compact set A. Then $x^0 = \sum_{k=1}^{r} \alpha_k^0 a^k$ is an optimal solution to Problem (I). Now, $g(x^0) = \sum_{k=1}^{r} \alpha_k^0 g(a^k) \geq \min_k g(a^k)$, thus $g(x^0) = \sum_{k=1}^{r} \alpha_k^0 g(a^k) = \min_k g(a^k)$ and every a^{k_0} such that $g(a^{k_0}) = \min_k g(a^k)$ is also an optimal solution to Problem (I). Moreover, in the expression $x^0 = \sum_{k=1}^{r} \alpha_k^0 a^k$ of every optimal solution x^0 in A to problem (I), one obviously has $\alpha_k^0 = 0$ if $g(a^k) > g(a^{k_0})$, which proves the last assertion. As proved in Proposition 3.2.3, if P does not contain any line, the vertices of A are the extreme points of P. □

Proposition 4.1.2, together with Proposition 4.1.1, indicates how to solve a linear programming problem as Problem (I). If $P = A + C$, the feasible set of (I), does not contain any line, it suffices to determine the extreme points of A (using Proposition 3.3.1 in Chapter 3, since they are the extreme points of P) and, for each of them, to verify, using Proposition 4.1.1, whether it is or not an optimal solution of P. This method, very useful if the extreme points of A are easy to find by geometric inspection (a simple geometric inspection can also allow to decide whether these points are or not optimal solutions), can lead to an intolerable amount of work if P is defined by a large number of linear inequalities and equalities. Then, an algorithm is needed to go from an extreme point to an other one, improving at each step the value of the objective function. It is on this idea that lies the simplex method to be explained later.

4.2 The duality theorem of linear programming

4.2.1 Canonical form of the duality theorem

Let A be a $(m \times n)$-matrix, $b \in \mathbb{R}^m$, $c \in \mathbb{R}^n$. Let us consider the following linear programming problem (P), here referred to as the *primal problem*,

$$(P) \qquad \begin{aligned} &\text{Minimize } c \cdot x \text{ subject to} \\ &Ax \geq b \\ &x \geq 0 \\ &x \in \mathbb{R}^n. \end{aligned}$$

and define its *dual problem* (D)

$$(D) \qquad \begin{aligned} &\text{Maximize } b \cdot p \text{ subject to} \\ &{}^t pA \leq {}^t c \\ &p \geq 0 \\ &p \in \mathbb{R}^m. \end{aligned}$$

Obviously, restating the problem (D) as a minimization problem for $-b \cdot p$ and constraints $-{}^t pA \geq -{}^t c$, $p \geq 0$, $p \in \mathbb{R}^m$, that is,

$$\begin{aligned} &\text{Minimize } \; - b \cdot p \text{ subject to} \\ &{}^t Ap \geq -{}^t c \\ &p \geq 0 \\ &p \in \mathbb{R}^m, \end{aligned}$$

and writing down the dual of the dual

$$\begin{aligned} &\text{Maximize } \; - c \cdot x \text{ subject to} \\ &{}^t x(-{}^t A) \leq -{}^t b \\ &x \geq 0 \\ &x \in \mathbb{R}^n, \end{aligned}$$

we see that we have returned to the given primal problem. For this reason, the linear programming problems (P) and (D) are referred to as *primal problem* and *dual problem* in *canonical form*. Each one of both problems is said to be the *dual* of the other one.

Lemma 4.2.1. *If x and p are respectively feasible solutions to problems (P) and (D), then $p \cdot b \leq c \cdot x$. If, moreover, $p \cdot b = c \cdot x$, then x and p are respectively optimal solutions to problems (P) and (D).*

Proof. Assume that x and p are feasible. Then

$$p \cdot b = {}^t pb \leq {}^t pAx = ({}^t pA) \cdot x \leq {}^t cx = c \cdot x.$$

Assume moreover that $p \cdot b = c \cdot x$. If x is not optimal, consider x' such that $Ax' \geq b$, $x' \geq 0$, $c \cdot x' < c \cdot x$. Then $c \cdot x' < p \cdot b$, which contradicts the feasibility of x' and p. Analogously, if p is not optimal, consider p' such that ${}^t p'A \leq {}^t c$, $p' \geq 0$, $b \cdot p' > b \cdot p$. Then $b \cdot p' > c \cdot x$, which contradicts the feasibility of x and p'. □

Lemma 4.2.2. *The system of linear inequalities*

$$\begin{cases} Ax - tb & \geq 0 \\ -{}^t pA + t({}^t c) & \geq 0 \\ p \cdot b - c \cdot x & \geq 0 \end{cases}$$

has a solution $(p, x, t) \in \mathbb{R}^m_+ \times \mathbb{R}^n_+ \times \mathbb{R}_+$ satisfying:

$$\begin{aligned} Ax - tb + p &\gg 0; \\ -{}^t pA + t({}^t c) + {}^t x &\gg 0; \\ p \cdot b - c \cdot x + t &> 0. \end{aligned}$$

Proof. It suffices to consider the skew-symmetric matrix

$$K = \begin{pmatrix} 0 & A & -b \\ -{}^t A & 0 & c \\ {}^t b & -{}^t c & 0 \end{pmatrix}$$

and to apply Corollary 2.3.6 to Minkowski-Farkas' lemma to the system

$$K \begin{pmatrix} p \\ x \\ t \end{pmatrix} \geq 0, \; p \geq 0, \; x \geq 0, \; t \geq 0.$$

□

Proposition 4.2.1. *Given the pair of dual linear programming problems (P) and (D), one of the two following alternatives holds:*

- either (P) and (D) have a couple $(\overline{x}, \overline{p})$ of optimal solutions satisfying: $\overline{p} \cdot b = c \cdot \overline{x}$ (obviously, the same relation $p \cdot b = c \cdot x$ holds then for every couple (x, p) of optimal solutions);

- or neither (P) nor (D) has an optimal solution and one of both feasible sets is empty.

Proof. Let (p, x, t) be as in Lemma 4.2.2. We will distinguish two cases.

First case: $t > 0$. Then $\overline{x} = (1/t)x$ and $\overline{p} = (1/t)p$ are respectively feasible solutions to (P) and (D) such that $\overline{p} \cdot b - c \cdot \overline{x} \geq 0$. We deduce from the first assertion of Lemma 4.2.1, that $\overline{p} \cdot b - c \cdot \overline{x} = 0$ and from the second assertion of Lemma 4.2.1, that $\overline{x}$ and $\overline{p}$ are optimal solutions to problems (P) and (D).

Second case: $t = 0$. Then, $Ax \geq 0$, ${}^tpA \leq 0$, $x \geq 0$, $p \geq 0$. If both feasible sets are nonempty, consider (x^0, p^0), a couple of feasible solutions to problems (P) and (D), that is such that: $Ax^0 \geq b$, $x^0 \geq 0$; ${}^tp^0A \leq {}^tc$, $p^0 \geq 0$. One deduces: ${}^tpb \leq {}^tpAx^0 \leq 0$ and ${}^tcx \geq {}^tp^0Ax \geq 0$, which contradicts the relation $p \cdot b > c \cdot x$ in Lemma 4.2.2. We have thus proved that at least one feasible set is empty.

Let us now show that if x^0 is a feasible solution of (P), then x^0 is not optimal. Indeed, for every $\lambda \geq 0$, $x^0 + \lambda x$ is a feasible solution to (P). It also follows from Lemma 4.2.2 that $c \cdot x < p \cdot b \leq {}^tpAx^0 \leq 0$, which implies $c \cdot (x^0 + \lambda x) < c \cdot x^0$. Thus the problem (P) has no optimal solution. Moreover, the quantity $c \cdot (x^0 + \lambda x)$ tends to $-\infty$ when λ tends to $+\infty$. The problem (P) is said to be *unbounded.*

With analogous arguments, one shows that if (D) has a feasible solution then (D) has no optimal solution and is unbounded. □

Corollary 4.2.1 (Complementarity relations). *In order that a couple $(\overline{x}, \overline{p})$ of feasible solutions to (P) and (D) is a couple of optimal solutions, it is necessary and sufficient that $({}^t\overline{p}A - {}^tc)\overline{x} = {}^t\overline{p}(A\overline{x} - b) = 0$, a condition which can be rewritten:*

$$({}^t\overline{p}A - {}^tc)_i \overline{x}_i = 0, \quad \forall i = 1, \dots, n \tag{4.2}$$

and

$${}^t\overline{p}_j (A\overline{x} - b)_j = 0, \quad \forall j = 1, \dots, m. \tag{4.3}$$

Proof. Let $(\overline{x}, \overline{p})$ be a couple of feasible solutions to (P) and (D). If $\overline{x}$ and $\overline{p}$ are optimal solutions, then the first alternative of the previous proposition holds and ${}^tc\overline{x} = {}^t\overline{p}b$. Using the feasibility relations, one gets:

$$0 = {}^t\overline{p}b - {}^tc\overline{x} \leq {}^t\overline{p}A\overline{x} - {}^tc\overline{x} = ({}^t\overline{p}A - {}^tc)\overline{x} \leq 0$$
$$0 = {}^t\overline{p}b - {}^tc\overline{x} \leq {}^t\overline{p}b - {}^t\overline{p}A\overline{x} = {}^t\overline{p}(A\overline{x} - b) \leq 0,$$

which shows that $({}^t\overline{p}A - {}^tc)\overline{x} = {}^t\overline{p}(A\overline{x} - b) = 0$.

Assume conversely that $({}^t\overline{p}A - {}^tc)\overline{x} = {}^t\overline{p}(A\overline{x} - b) = 0$. Then $c \cdot \overline{x} = {}^tc\overline{x} = {}^t\overline{p}A\overline{x} = {}^t\overline{p}b = p \cdot b$, and it follows from Lemma 4.2.1 that $\overline{x}$ and $\overline{p}$ are optimal solutions to respectively (P) and (D). □

4.2.2 General form of the duality theorem

Linear programming problems are not necessarily given in canonical form. In order to state the duality theorem for these problems, we suggest in this subsection several procedures for putting a linear programming problem into canonical form. Let A be a $(m \times n)$-matrix, A' be a $(p \times n)$-matrix and let $b \in \mathbb{R}^m$, $b' \in \mathbb{R}^p$, $c \in \mathbb{R}^n$. We now consider a linear programming problem given in the following more general form:

$$(P') \qquad \begin{array}{ll} \multicolumn{2}{c}{\text{Minimize } c \cdot x \text{ subject to}} \\ Ax \geq b & (m \text{ linear inequalities}) \\ A'x = b' & (p \text{ linear equalities}) \\ x \in \mathbb{R}^n. & \end{array}$$

This problem can be rephrased in the following successive forms:

$$\text{Minimize } c \cdot x \text{ subject to}$$

$$\begin{array}{c} Ax \geq b \\ A'x \geq b' \\ -A'x \geq -b' \\ x \in \mathbb{R}^n \end{array} \qquad \text{that is} \qquad \begin{array}{c} \begin{pmatrix} A \\ A' \\ -A' \end{pmatrix} x \geq \begin{pmatrix} b \\ b' \\ -b' \end{pmatrix} \\ x \in \mathbb{R}^n \end{array}$$

then, setting $x = x^+ - x^-$, where $x^+ \geq 0$ and $x^- \geq 0$ are respectively the positive and the negative part in $\mathbb{R}^n$ of x,

$$\begin{array}{c} \text{Minimize } \begin{pmatrix} c \\ -c \end{pmatrix} \cdot \begin{pmatrix} x^+ \\ x^- \end{pmatrix} \text{ subject to} \\ \begin{pmatrix} A & -A \\ A' & -A' \\ -A' & A' \end{pmatrix} \begin{pmatrix} x^+ \\ x^- \end{pmatrix} \geq \begin{pmatrix} b \\ b' \\ -b' \end{pmatrix} \\ x^+ \geq 0, \; x^- \geq 0 \\ \begin{pmatrix} x^+ \\ x^- \end{pmatrix} \in \mathbb{R}^{2n}. \end{array}$$

According to the definition given above, the dual of the previous problem is:

$$\begin{array}{c} \text{Maximize } \begin{pmatrix} b \\ b' \\ -b' \end{pmatrix} \cdot \begin{pmatrix} p \\ r \\ s \end{pmatrix} \text{ subject to} \\ ({}^t p \quad {}^t r \quad {}^t s) \begin{pmatrix} A & -A \\ A' & -A' \\ -A' & A' \end{pmatrix} \geq \begin{pmatrix} c \\ -c \end{pmatrix} \\ p \geq 0, \; r \geq 0, \; s \geq 0 \\ p \in \mathbb{R}^m, \; r \in \mathbb{R}^p, \; s \in \mathbb{R}^p \end{array}$$

equivalently, setting $q = r - s$,

$$(D') \qquad \begin{array}{c} \text{Maximize } \begin{pmatrix} b \\ b' \end{pmatrix} \cdot \begin{pmatrix} p \\ q \end{pmatrix} = b \cdot p + b' \cdot q \text{ subject to} \\ {}^t pA + {}^t qA' = c, \text{ that is, } ({}^t p \quad {}^t q) \begin{pmatrix} A \\ A' \end{pmatrix} = c \\ p \geq 0 \\ p \in \mathbb{R}^m, \; q \in \mathbb{R}^p. \end{array}$$

More generally again, let A be a $(m \times n)$-matrix, A' be a $(p \times n)$-matrix, $b \in \mathbb{R}^m$, $b' \in \mathbb{R}^p$, $c \in \mathbb{R}^n$ and let J be a subset of the set of indices $\{1, \dots, n\}$. Consider the problem:

$$(P'') \qquad \begin{array}{l} \text{Minimize } c \cdot x \text{ subject to} \\ \quad Ax \geq b \\ \quad A'x = b' \\ \quad x_j \geq 0, \ \forall j \in J \\ \quad x \in \mathbb{R}^n. \end{array}$$

Let us denote by a^j (resp. a'^j) the jth column-vector of the matrix A (resp. A'). One can verify that the dual problem (D'') of (P'') is:

$$(D'') \qquad \begin{array}{l} \text{Maximize } \begin{pmatrix} b \\ b' \end{pmatrix} \cdot \begin{pmatrix} p \\ q \end{pmatrix} \text{ subject to} \\ {}^t p a^j + {}^t q a'^j \leq c_j, \ \forall j \in J \\ {}^t p a^j + {}^t q a'^j = c_j, \ \forall j \notin J \\ p \geq 0 \\ p \in \mathbb{R}^m, \ q \in \mathbb{R}^p. \end{array}$$

From the last example, one sees that the rules for passing from a linear programming problem to its dual are:

- to a minimization (resp. maximization) problem corresponds a maximization (resp. minimization) problem;
- to inequalities in the primal problem correspond positivity constraints for the variables of the dual problem;
- to equalities in the primal problem correspond variables of the dual problem without restrictions as to sign;
- to positivity constraints for the variables of the primal problem correspond inequalities in the dual problem;
- the matrix whose elements are the coefficients of the "true" constraints (the linear constraints other than positivity restrictions) is the same in both problems; if it is multiplied (resp. pre-multiplied) by the column vector (resp. row vector) of the primal variables, it is pre-multiplied (resp. multiplied) by the row vector (resp. column vector) of the dual variables;
- to the vector of the right members of the true constraints in each problem correspond the coefficients of the linear functional to optimize in the other one.

The general form of the duality theorem and of the complementarity relations are the following:

Proposition 4.2.2. *Given the pair of linear programming problems (P'') and (D''), one of the two following alternatives holds:*

- either (P'') and (D'') have a couple $(\overline{x}, (\overline{p}, \overline{q}))$ of optimal solutions satisfying: $b \cdot \overline{p} + b' \cdot \overline{q} = c \cdot \overline{x}$ (both problems have the same optimal value)

- or neither (P'') nor (D'') has an optimal solution and one of both feasible sets is empty.

Proof. The proof is similar to the proof of Proposition 4.2.1. □

Corollary 4.2.2. *In order that a couple* $(\overline{x}, (\overline{p}, \overline{q}))$ *of feasible solutions to* (P'') *and* (D'') *is a couple of optimal solutions, it is necessary and sufficient that*

$$\overline{x}_j({}^t pa^j + {}^t qa'^j - c_j) = 0, \ \forall j \in J \tag{4.4}$$

and

$$\overline{p}_k(b - A\overline{x})_k = 0, \ \forall k = 1, \ldots, m. \tag{4.5}$$

Proof. The proof is similar to the proof of Corollary 4.2.1. □

Duality is used in linear programming in the following way. Faced with a problem, it may appear that the dual problem is easier to analyze. The duality theorem allows then to decide whether or not the initial problem has a solution and to compute (if it exists) this solution, via the complementarity relations.
Besides, duality arguments are involved in the proof of several results. In what follows, the duality theorem is used to establish an important result on linear inequalities which extends the proposition 2.3.1 of Chapter 2.

4.2.3 Application to nonhomogeneous Farkas' lemma

Proposition 4.2.3. *Let for every* $i = 0, 1, \ldots, m$, $a^i \in \mathbb{R}^n$, $\alpha_i \in \mathbb{R}$ *be such that the system* $a^i \cdot x \leq \alpha_i$, $i = 1, \ldots, m$ *is consistent. Then in order that the following implication is true*

$$a^i \cdot x \leq \alpha_i, \ i = 1, \ldots, m \Rightarrow a^0 \cdot x \leq \alpha_0 \tag{4.6}$$

it is necessary and sufficient that there exist nonnegative real numbers $\lambda_i \geq 0$ *such that*

$$a^0 = \sum_{i=1}^m \lambda_i a^i \text{ and } \alpha_0 \geq \sum_{i=1}^m \lambda_i \alpha_i.$$

Proof. The condition is obviously sufficient. We now prove that it is necessary. Let us first remark that the following implication is true

$$a^i \cdot y \leq 0, \ i = 1, \ldots, m \Rightarrow a^0 \cdot y \leq 0. \tag{4.7}$$

Indeed if some y satisfies $a^i \cdot y \leq 0$, $i = 1, \ldots, m$ and if x^0 is some solution to the system $a^i \cdot x \leq \alpha_i$, $i = 1, \ldots, m$, we have then for every $\lambda > 0$, $a^i \cdot (x^0 + \lambda y) \leq a^i \cdot x^0 \leq \alpha_i$, $i = 1, \ldots, m$ thus $a^i \cdot (x^0 + \lambda y) \leq \alpha_i$, $i = 1, \ldots, m$ and, in view of (4.6), $a^0 \cdot (x^0 + \lambda y) \leq \alpha_0$. Dividing by $\lambda > 0$ and letting λ tend to $+\infty$, we get $a^0 \cdot y \leq 0$, which proves (4.7).
From Minkowski-Farkas' lemma (Proposition 2.3.1), one then deduces

$$\exists \mu_i \geq 0, \ i = 1, \ldots, m, \ a^0 = \sum_{i=1}^m \mu_i a^i.$$

Let us now consider the linear programming problem:

$$(P) \qquad \begin{array}{l} \text{Maximize } a^0 \cdot x \text{ subject to} \\ a^i \cdot x \leq \alpha_i, \ i = 1, \ldots, m, \\ x \in \mathbb{R}^n \end{array}$$

whose dual is:

$$(D) \qquad \begin{array}{l} \text{Minimize } \sum_{i=1}^m \mu_i \alpha_i \text{ subject to} \\ \sum_{i=1}^m \mu_i a^i = a^0, \\ \mu = (\mu_i) \in \mathbb{R}_+^m. \end{array}$$

We have just proved above that the problem (D) has a feasible solution. We know from the hypothesis that the feasible set of the problem (P) is also nonempty. In view of Proposition 4.2.2 and Corollary 4.2.2, there exists a couple $(\overline{x}, \lambda)$ of optimal solutions respectively to (P) and (D) such that $\lambda_i(a^i \cdot \overline{x} - \alpha_i) = 0, \forall i = 1, \ldots, m$. In view of (4.6) together with the feasibility conditions, the last relations imply that $a^0 = \sum_{i=1}^m \lambda_i a^i$ and

$$\sum_{i=1}^m \lambda_i \alpha_i = \sum_{i=1}^m \lambda_i a^i \cdot \overline{x} = a^0 \cdot \overline{x} \leq \alpha_0. \qquad \square$$

4.3 The simplex method

4.3.1 Extreme points of the set of nonnegative solutions to a system of linear equations

Let $P = \{x \in \mathbb{R}_+^{n+m} \mid Ax = b\}$ where A is a $(m \times (n+m))$-matrix of rank m and $b \in \mathbb{R}^m$. We begin with a result which sharpens the proposition 3.3.1 established in Chapter 3.

Proposition 4.3.1. *If $\{a^{i_1}, \ldots, a^{i_m}\}$ is a system of m linearly independent column vectors of A, any nonnegative solution to*

$$\begin{pmatrix} a^{i_1} & \ldots & a^{i_m} \end{pmatrix} \begin{pmatrix} x_{i_1} \\ \vdots \\ x_{i_m} \end{pmatrix} = b$$

determines an extreme point $(0, \ldots, x_{i_1}, \ldots, x_{i_m}, \ldots, 0)$ of the set P in which at least n coordinates are 0. Conversely, any extreme point of P is so determined for some selection of m linearly independent column vectors of A.

Proof. In view of Proposition 3.3.1, we have only to prove the second assertion. Let x be an extreme point of P and let $\{i_1, \ldots, i_p\}$ be the set of indices corresponding to positives entries. One has:

$$\begin{pmatrix} a^{i_1} & \dots & a^{i_p} \end{pmatrix} \begin{pmatrix} x_{i_1} \\ \vdots \\ x_{i_p} \end{pmatrix} = b.$$

We first claim that the set $\{a^{i_1}, \dots, a^{i_p}\}$ is linearly independent, which implies $p \leq m$. Indeed, if not, the system $\begin{pmatrix} a^{i_1} & \dots & a^{i_p} \end{pmatrix} \begin{pmatrix} z_{i_1} \\ \vdots \\ z_{i_p} \end{pmatrix} = 0$ would have a solution such that $z = (0, \dots, z_{i_1}, \dots, z_{i_p}, \dots, 0) \neq 0$. Then for small enough λ, $x + \lambda z \in P$, $x - \lambda z \in P$, $x = (1/2)(x + \lambda z) + (1/2)(x - \lambda z)$ would imply that x is not be an extreme point of P.

Since $p \leq m$ and since the matrix A has rank m, one can complete as necessary the system $\{a^{i_1}, \dots, a^{i_p}\}$ with column vectors of A so as to get a system of linearly independent column vectors $\{a^{i_1}, \dots, a^{i_p}, a^{i_{p+1}}, \dots, a^{i_m}\}$. Then

$$\begin{pmatrix} a^{i_1} & \dots & a^{i_p} & a^{i_{p+1}} & \dots & a^{i_m} \end{pmatrix} \begin{pmatrix} x_{i_1} \\ \vdots \\ x_{i_p} \\ 0 \\ \vdots \\ 0 \end{pmatrix} = b.$$

□

4.3.2 Solving a linear programming problem in canonical form

For a $(m \times n)$-matrix $A = (a_{ij})_{\substack{i=1,\dots,m \\ j=1,\dots,n}}$, $b \in \mathbb{R}^m$, $c \in \mathbb{R}^n$, $d \in \mathbb{R}$, let us now consider a linear programming problem in canonical form

$$(P) \qquad \begin{aligned} &\text{Minimize } f(x) = c \cdot x + d \text{ subject to} \\ &Ax \geq b \\ &x \geq 0 \\ &x \in \mathbb{R}^n. \end{aligned}$$

Introducing additional variables $x_{n+1}, \dots, x_{n+m}$, called *slack variables*, it is convenient to associate with (P) the problem in *standard form*:

$$(P') \qquad \begin{aligned} &\text{Minimize } f'(x') = c' \cdot x' + d \text{ subject to} \\ &A'x' = b \\ &x' \geq 0 \\ &x' \in \mathbb{R}^{n+m}. \end{aligned}$$

where $A' = \begin{pmatrix} A & -I_D \end{pmatrix}$ (I_D is the $(m \times m)$-identity matrix) and $c' = (c, 0, \dots, 0)$.

Let P and P' be the feasible sets of respectively (P) and (P'). One immediately sees that $x \in P$ if and only if there exists $x_{n+1}, \dots, x_{n+m}$ such that $x' = (x_1, \dots, x_n, x_{n+1}, \dots, x_{n+m}) \in P'$. Likewise, x is an optimal solution to (P) if and only if x' is an optimal solution to (P'). A possible procedure for solving (P) consists in solving (P'), then in taking, when (P') has an optimal solution, the projection of solutions to (P') onto the linear subspace of $\mathbb{R}^{n+m}$ generated by the first n vectors of the canonical basis of $\mathbb{R}^{n+m}$.

Moreover, using Proposition 4.3.1, we see that, under the condition $b_i \leq 0$, $\forall i = 1, \dots, m$, the point $(0, \dots, 0, -b_1, \dots, -b_m)$ is an extreme point of the feasible P' of (P'). It is an optimal solution to (P') (with d as optimal value) under the additional condition $c_j \geq 0$, $\forall j = 1, \dots, n$. This optimal solution to (P') is unique if $c_j > 0$, $\forall j = 1, \dots, n$.

With these remarks in mind, the simplex method proposes successive transformations of the variables and the constants of problem (P'), so as to get, in a reasonable number of steps, a new problem equivalent to (P') and expressed in the same form, but with constants $\overline{b} = (\overline{b}_1, \dots, \overline{b}_m)$ and $\overline{c} = (\overline{c}_1, \dots, \overline{c}_n)$ of the required sign: $\overline{b}_i \leq 0$, $\forall i = 1, \dots, m$, $\overline{c}_j \geq 0$, $\forall j = 1, \dots, n$.

4.3.3 Pivoting

Let us call a *tableau* corresponding to Problem (P') the following array:

$$\begin{array}{c} \\ x_{n+1} \\ \\ x_{n+i} \\ \\ x_{n+m} \\ f' \end{array} \begin{array}{c} \begin{array}{ccccccc} x_1 & & x_j & & x_n & & \end{array} \\ \begin{pmatrix} a_{11} & \dots & a_{1j} & \dots & a_{1n} & -b_1 \\ \vdots & & \vdots & & \vdots & \\ a_{i1} & \dots & a_{ij} & \dots & a_{in} & -b_i \\ \vdots & & \vdots & & \vdots & \\ a_{m1} & \dots & a_{mj} & \dots & a_{mn} & -b_m \\ c_1 & \dots & c_j & \dots & c_n & d \end{pmatrix} \end{array} \tag{4.8}$$

which summarizes the data of the problem. If $a_{ij} \neq 0$, it is easily seen that the constraints which define P'

$$\begin{array}{ccccccc} a_{11}x_1 & + \dots & +a_{1j}x_j & + \dots & +a_{1n}x_n & -b_1 & = x_{n+1} \\ \vdots & & \vdots & & \vdots & \vdots & \vdots \\ a_{i1}x_1 & + \dots & +a_{ij}x_j & + \dots & +a_{in}x_n & -b_i & = x_{n+i} \\ \vdots & & \vdots & & \vdots & \vdots & \vdots \\ a_{m1}x_1 & + \dots & +a_{mj}x_j & + \dots & +a_{mn}x_n & -b_m & = x_{n+m} \end{array}$$

together with the restrictions $x_k \geq 0$, $k = 1, \dots, n+m$, are equivalent to

$$
\begin{array}{llll}
(a_{11} - \frac{a_{1j}}{a_{ij}} a_{i1})x_1 & + \ldots + \frac{a_{1j}}{a_{ij}} x_{n+j} & + \ldots + (a_{1n} - \frac{a_{1j}}{a_{ij}} a_{in})x_n & \\
 & & -(b_1 - \frac{a_{1j}}{a_{ij}} b_i) = x_{n+1} & \\
\vdots & \vdots & \vdots & \\
 & & & \vdots \\
(-\frac{a_{i1}}{a_{ij}})x_1 & + \ldots + \frac{1}{a_{ij}} x_{n+j} & + \ldots + (-\frac{a_{in}}{aij})x_n & \\
 & & -(-\frac{b_i}{a_{ij}}) = x_j & \\
\vdots & \vdots & \vdots & \\
 & & & \vdots \\
(a_{m1} - \frac{a_{mj}}{a_{ij}} a_{i1})x_1 & + \ldots + \frac{a_{mj}}{a_{ij}} x_{n+j} & + \ldots + (a_{mn} - \frac{a_{mj}}{a_{ij}} a_{in})x_n & \\
 & & -(b_m - \frac{a_{mj}}{a_{ij}} b_i) = x_{n+m} &
\end{array}
$$

with the same positivity restrictions. In this system, the variables

$$(x_{n+1}, \ldots, x_{n+j-1}, x_j, x_{n+j+1}, \ldots, x_{n+m})$$

are now expressed as functions of

$$(x_1, \ldots, x_{j-1}, x_{n+j}, x_{j+1}, \ldots, x_n).$$

The function to minimize is

$$f'(x') = (c_1 - \frac{a_{i1}}{a_{ij}} c_j)x_1 + \ldots + \frac{c_j}{a_{ij}} x_{n+j} + \ldots + (c_n - \frac{a_{in}}{a_{ij}} c_j)x_n + c_j \frac{b_i}{a_{ij}} + d$$

and the new tableau (4.9)

$$
\begin{array}{c}
 \\ x_{n+1} \\ \\ x_j \\ \\ x_{n+m} \\ f'
\end{array}
\begin{array}{c}
\begin{array}{ccccccc} x_1 & & x_{n+j} & & x_n & & \end{array} \\
\begin{pmatrix}
a_{11} - \frac{a_{1j}}{a_{ij}} a_{i1} & \ldots & \frac{a_{1j}}{a_{ij}} & \ldots & a_{1n} - \frac{a_{1j}}{a_{ij}} a_{in} & -(b_1 - \frac{a_{1j}}{a_{ij}} b_i) \\
\vdots & & \vdots & & \vdots & \\
-\frac{a_{i1}}{a_{ij}} & \ldots & \frac{1}{a_{ij}} & \ldots & -\frac{a_{in}}{aij} & -\frac{-b_i}{a_{ij}} \\
\vdots & & \vdots & & \vdots & \\
a_{m1} - \frac{a_{mj}}{a_{ij}} a_{i1} & \ldots & \frac{a_{mj}}{a_{ij}} & \ldots & a_{mn} - \frac{a_{mj}}{a_{ij}} a_{in} & -(b_m - \frac{a_{mj}}{a_{ij}} b_i) \\
c_1 - \frac{a_{i1}}{a_{ij}} c_j & \ldots & \frac{c_j}{a_{ij}} & \ldots & c_n - \frac{a_{in}}{a_{ij}} c_j & c_j \frac{b_i}{a_{ij}} + d
\end{pmatrix}
\end{array}
\tag{4.9}
$$

summarizes the system. The process of going from tableau (4.8) to (4.9) is called *pivoting about* a_{ij}.

The simplex method defines several pivoting steps in order to get a final tableau

$$\begin{array}{c} \\ x_{\overline{n+1}} \\ \\ x_{\overline{n+i}} \\ \\ x_{\overline{n+m}} \\ f' \end{array} \begin{pmatrix} x_{\overline{1}} & & x_{\overline{j}} & & x_{\overline{n}} & \\ \overline{a}_{11} & \ldots & \overline{a}_{1j} & \ldots & \overline{a}_{1n} & -\overline{b}_1 \\ \vdots & & \vdots & & \vdots & \\ \overline{a}_{i1} & \ldots & \overline{a}_{ij} & \ldots & \overline{a}_{in} & -\overline{b}_i \\ \vdots & & \vdots & & \vdots & \\ \overline{a}_{m1} & \ldots & \overline{a}_{mj} & \ldots & \overline{a}_{mn} & -\overline{b}_m \\ \overline{c}_1 & \ldots & \overline{c}_j & \ldots & \overline{c}_n & \overline{d} \end{pmatrix} \tag{4.10}$$

where the constants $\overline{b}_i$, $i = 1, \ldots m$ and $\overline{c}_j$, $j = 1, \ldots, n$ have the required sign: $\overline{b}_i \leq 0$ and $\overline{c}_j \geq 0$. Observe that, in this formulation of Problem (P'), $\overline{a}_{ij}, \overline{b}_i, \overline{c}_j$ denote the new values of coefficients, while $\{\overline{1},..., \overline{n}, \overline{n+1},...,\overline{n+m}\}$ is some permutation of $\{1, \ldots, n, n+1, \ldots, n+m\}$.

The method is generally described in two phases. In Phase I, the purpose is to get all constants $\overline{b}_i \leq 0$ and so to determine an extreme point of the feasible set of Problem (P'). In phase II, the purpose is to get all coefficients $\overline{c}_j \geq 0$. We first describe this second phase.

Phase II (getting all $c_j \geq 0$). Let us begin by assuming that we have a tableau like (4.8) in which for all i, $b_i \leq 0$. Recall that, in view of proposition 4.3.1, it implies that the point $(0, \ldots, 0, -b_1, \ldots, -b_m)$ is an extreme point of the feasible set. The value at this point of the objective function is $f'(0, \ldots, 0, -b_1, \ldots, -b_m) = d$. Let j be such that $c_j < 0$.
If $a_{ij} \geq 0$, $\forall i = 1, \ldots, m$, then for every $\lambda > 0$, the point

$$(0, \ldots, \lambda, \ldots, 0, a_{1j}\lambda - b_1, \ldots, a_{mj}\lambda - b_m)$$

is a feasible solution, so that the objective function of Problem (P') is unbounded below on the feasible set and this problem has no solution.
If not, among all entries in column j with $a_{kj} < 0$, choose i such that $(b_i/a_{ij}) \leq (b_k/a_{kj})$. As it is easily verified, pivoting about a_{ij} leads to a new tableau such that $\overline{c}_j > 0$ and $\overline{b}_k \leq 0$, $\forall k = 1, \ldots, m$, consequently to a new extreme point of the feasible set defined by:

$$x_1 = \ldots = x_{n+j} = \ldots = 0, \ x_{n+1} = -\overline{b}_1, \ldots, x_j = -\overline{b}_j, \ldots, x_{n+m} = -\overline{b}_m.$$

The value of the objective function at this point is equal to: $c_j(b_i/a_{ij}) + d$, thus inferior or equal to its value, d, at the previous extreme point.

Following the rules outlined above, one applies a sequence of pivots until to get a system with coefficients such that:

$$\overline{\overline{b}}_k \leq 0, \ \forall k = 1, \ldots, m; \ \ \overline{\overline{c}}_j \geq 0, \ \forall j = 1, \ldots, n.$$

The point

$$(x_{\overline{\overline{1}}}, \ldots, x_{\overline{\overline{n}}}, x_{\overline{\overline{n+1}}}, \ldots, x_{\overline{\overline{n+m}}}) = (0, \ldots, 0, -b_{\overline{\overline{1}}}, \ldots, -b_{\overline{\overline{m}}})$$

is an optimal solution to Problem (P'). As $\{\bar{\bar{1}},\dots,\bar{\bar{n}},\dots,\overline{\overline{n+m}}\}$ is a permutation of $\{1,\dots,n+m\}$, this point corresponds to some $(x_1,\dots,x_n,\dots,x_{n+m})$ such that $(x_1,\dots,x_n)$ is an optimal solution to Problem (P).

It is worth noticing that three cases can occur following the above procedure:

- the sequence of pivots leads to an optimal solution to the linear programming problem in a finite number of steps;
- the sequence of pivots leads to constants such that $\bar{\bar{c}}_j < 0$ for some j together with $\bar{\bar{a}}_{ij} \geq 0$, $\forall i = 1,\dots,m$. In this case, the problem has no solution;
- the algorithm (which consists in going from an extreme point of the feasible set to another one, getting each time a value of the objective function smaller than or equal to its value at the previous one) runs into an infinite sequence of pivots. This phenomenon, called *cycling*, prevents to find a solution. More sophisticated simplex methods, that we will not describe here, allow to cope with this problem.

Phase I (getting all $b_i \leq 0$). We now consider a problem where at least one $b_i > 0$. For solving this problem, let us first introduce an additional variable x_0, set $x'' = (x_0, x_1, \dots, x_{n+m})$ and consider the problem:

$$(P'')\quad \begin{array}{c} \text{Minimize } g(x'') = x_0 \text{ subject to} \\ \begin{array}{ccccccccc} x_0 & +a_{11}x_1 & +\dots & +a_{1j}x_j & +\dots & +a_{1n}x_n & -b_1 & = x_{n+1} \\ \vdots & \vdots & & \vdots & & \vdots & \vdots & \vdots \\ x_0 & +a_{i1}x_1 & +\dots & +a_{ij}x_j & +\dots & +a_{in}x_n & -b_i & = x_{n+i} \\ \vdots & \vdots & & \vdots & & \vdots & \vdots & \vdots \\ x_0 & +a_{m1}x_1 & +\dots & +a_{mj}x_j & +\dots & +a_{mn}x_n & -b_m & = x_{n+m} \end{array} \\ x_0 \geq 0,\ x_1 \geq 0, \dots,\ x_{n+m} \geq 0 \end{array}$$

with tableau

$$\begin{array}{c} \\ x_{n+1} \\ \\ x_{n+i} \\ \\ x_{n+m} \\ g \end{array} \begin{array}{c} \begin{array}{ccccc} x_0 & x_1 & & x_n & \end{array} \\ \left(\begin{array}{ccccc} 1 & a_{11} & \dots & a_{1n} & -b_1 \\ \vdots & \vdots & & \vdots & \\ 1 & a_{i1} & \dots & a_{in} & -b_i \\ \vdots & \vdots & & \vdots & \\ 1 & a_{m1} & \dots & a_{mn} & -b_m \\ 1 & 0 & \dots & 0 & 0 \end{array} \right) \end{array} \tag{4.11}$$

Note first that the feasible set of (P'') is nonempty. It suffices to choose $x_0 \geq \max\{b_1,\dots,b_m\}$, $x_1 = \dots = x_n = 0$ and to determine x_{n+1}, ..., x_{n+m}. Since $x_0 \geq 0$, it follows from Proposition 4.1.2 that the problem (P'') has an optimal solution which is at least 0. Moreover, the constraints of (P'') reduces to the constraints of (P') when $x_0 = 0$. Consequently, the original constraints are satisfied if and only if the minimum value of $g(x'')$ is equal to 0. In particular, the feasible set of Problem (P') is empty if the optimal value of Problem (P'') is more than 0.

Let i be such that $b_i \geq b_k$, $\forall k = 1, \ldots, m$. Observe that $b_i > 0$. Pivoting about a_{i0} leads to the following new tableau for (P''):

$$\begin{array}{c} \\ x_{n+1} \\ \\ x_0 \\ \\ x_{n+m} \\ g \end{array} \begin{array}{c} \begin{array}{cccccc} x_{n+i} & x_1 & & x_n & \end{array} \\ \left(\begin{array}{cccccc} 1 & a_{11} - a_{i1} & \ldots & a_{1n} - a_{in} & -(b_1 - b_i) \\ \vdots & \vdots & & \vdots & \\ 1 & -a_{i1} & \ldots & -a_{in} & -(-b_i) \\ \vdots & \vdots & & \vdots & \\ 1 & a_{m1} - a_{i1} & \ldots & a_{mn} - a_{in} & -(b_m - b_i) \\ 1 & -a_{i1} & \ldots & -a_{in} & 0 \end{array}\right) \end{array} \tag{4.12}$$

in which all the $\overline{b}_i \leq 0$. The solution of (P'') may now be completed using the procedure described in Phase II. If the algorithm does not cycle, it leads to a final tableau:

$$\begin{array}{c} \\ x_{\overline{\overline{n+1}}} \\ \\ x_{\overline{\overline{n+m}}} \\ g \end{array} \begin{array}{c} \begin{array}{ccccc} x_{\overline{\overline{0}}} & x_{\overline{\overline{1}}} & & x_{\overline{\overline{n}}} & \end{array} \\ \left(\begin{array}{ccccc} \overline{\overline{a}}_{10} & \overline{\overline{a}}_{11} & \ldots & \overline{\overline{a}}_{1n} & -\overline{\overline{b}}_1 \\ \vdots & \vdots & & \vdots & \\ \overline{\overline{a}}_{m0} & \overline{\overline{a}}_{m1} & \ldots & \overline{\overline{a}}_{mn} & -\overline{\overline{b}}_m \\ \overline{\overline{c}}_0 & \overline{\overline{c}}_1 & \ldots & \overline{\overline{c}}_n & \overline{\overline{d}} \end{array}\right) \end{array} \tag{4.13}$$

where $\overline{\overline{d}}$ is the optimal value of (P'').

If $\overline{\overline{d}} > 0$, as already observed, the feasible set of problem (P') is empty.

If $\overline{\overline{d}} = 0$, recalling that the set of indices $\{\overline{\overline{0}}, \overline{\overline{1}}, \ldots, \overline{\overline{n+m}}\}$ is a permutation of $\{0, 1, \ldots, n+m\}$, one can assume without loss of generality that x_0 belongs to the set $\{x_{\overline{\overline{0}}}, x_{\overline{\overline{1}}}, \ldots, x_{\overline{\overline{n}}}\}$ (if 0 corresponds to $\overline{\overline{n+k}}$, then $\overline{\overline{b}}_k = 0$ and pivoting about any nonzero entry in the corresponding row will bring $x_{\overline{\overline{n+k}}}$, that is x_0, at the top of the tableau). Assume that $0 = \overline{\overline{j}}$. Deleting column $\overline{\overline{j}}$ from Tableau 4.13, we see that the relations

$$\begin{array}{ccccccc} \overline{\overline{a}}_{10} x_{\overline{\overline{0}}} & + \ldots + \overline{\overline{a}}_{1(j-1)} x_{\overline{\overline{j-1}}} + \overline{\overline{a}}_{1(j+1)} x_{\overline{\overline{j+1}}} & + \ldots + \overline{\overline{a}}_{1n} x_{\overline{\overline{n}}} & -\overline{\overline{b}}_1 = x_{\overline{\overline{n+1}}} \\ \vdots & \vdots & \vdots & \vdots \\ \overline{\overline{a}}_{m0} x_{\overline{\overline{0}}} & + \ldots + \overline{\overline{a}}_{m(j-1)} x_{\overline{\overline{j-1}}} + \overline{\overline{a}}_{m(j+1)} x_{\overline{\overline{j+1}}} & + \ldots + \overline{\overline{a}}_{mn} x_{\overline{\overline{n}}} & -\overline{\overline{b}}_m = x_{\overline{\overline{n+1}}} \end{array}$$

together with the positivity restrictions define the feasible set of Problem (P'). After expressing the objective function of this problem in terms of the permuted coordinates, we may proceed with Phase II to obtain an optimal solution to Problem (P'), and finally an optimal solution to Problem (P).

4.3.4 Practicing the simplex method

Let us consider the linear programming problem:

Minimize the affine functional $f(x) = \sum_{j=1}^{n} c_j x_j + d$ under constraints which can be expressed after adding slack variables:

$$\begin{array}{cccccccc} a_{11}x_1 & +\dots & +a_{1j}x_j & +\dots & +a_{1n}x_n & -b_1 & = x_{n+1} \\ \vdots & & \vdots & & \vdots & \vdots & \vdots \\ a_{m1}x_1 & +\dots & +a_{mj}x_j & +\dots & +a_{mn}x_n & -b_m & = x_{n+m} \end{array}$$

together with the restrictions $x_k \geq 0,\ k = 1, \dots, n+m$. The previous results can be summarized in the following instructions:

Phase I (Getting all $b_i \leq 0$)
1. If some $b_i > 0$ in the linear programming problem, set up the tableau (4.11) corresponding with the additional problem (P'').
2. Select i so that $b_i \geq b_k,\ \forall k = 1, \dots, m$.
3. Use a_{i0} ($a_{i0} = 1$) as a pivot to obtain a tableau (4.12)
4. Use Phase II (below) to solve (P'') and to get a final tableau (4.13).
5. If $\overline{\overline{d}} > 0$, STOP. The feasible set of the original problem is empty.
6. If $\overline{\overline{d}} = 0$ and if x_0 is at the top of (4.13), go to step 8.
7. If x_0 is some $x_{\overline{\overline{n+k}}}$, use any nonzero entry in the corresponding row as a pivot in order to bring x_0 at the top of the tableau.
8. Delete the column headed by x_0 and replace the last row by the constants of the objective function of the original problem, $f(x)$, expressed as a function of the remaining variables at the top of the tableau.
9. You can now go to Phase II for solving Problem (P').

Phase II (Getting all $c_j \geq 0$)
1. Begin with a tableau (4.8) such that $b_i \leq 0,\ \forall i = 1, \dots, m$.
2. If $\forall j = 1, \dots, n,\ c_j \geq 0$, STOP. The point $(0, \dots, 0, -b_1, , \dots, -b_m)$ is an optimal solution. Moreover, this solution is unique if $\forall j = 1, \dots, n,\ c_j > 0$.
3. Otherwise, select j so that $c_j < 0$.
4. If all entries in column j (except for c_j) are nonnegative, STOP. No optimal solution exists.
5. Otherwise, among those k with $a_{kj} < 0$, select i so that $b_i/a_{ij} \leq b_k/a_{kj}$.
6. Use a_{ij} as pivot to obtain a new tableau.
7. Go back to step 1.

Exercises

Exercise 4.1. Consider the problem:
Maximize $f(x_1, x_2) = -3x_1 + 2x_2$ subject to

$$\begin{array}{rcrcr} 2x_1 & - & x_2 & \geq & -2 \\ x_1 & - & 2x_2 & \leq & 3 \\ x_1 & + & 2x_2 & \leq & 11 \\ x_1 & \geq & 0, \ x_2 & \geq & 0. \end{array}$$

a) Represent the feasible set P.
b) Determine by geometric inspection the extreme points and extreme directions of P.
c) Express P as the sum $P = A + C$ of a polytope A and a polyhedral convex cone C.
d) Use the necessary and sufficient conditions of optimality for solving the problem.

Exercise 4.2. Same questions for the problem:
Maximize $f(x_1, x_2) = -x_1 + 4x_2$ subject to

$$\begin{array}{rcrcr} 2x_1 & - & x_2 & \geq & 4 \\ x_1 & - & 2x_2 & \geq & 3 \\ x_1 & \geq & 0, \ x_2 & \geq & 0. \end{array}$$

Exercise 4.3. Minimize $g(x_1, x_2) = x_1 + 4x_2$ subject to the constraints of Exercise 4.2.

Exercise 4.4. Set the dual problem of the problem considered in Exercise 4.1. Solve this problem and verify the duality theorem and the complementarity relations.

Exercise 4.5. What about the dual problem of the problem considered in Exercise 4.2?

Exercise 4.6. Consider the problem (P):
Minimize $f(x_1, x_2, x_3) = 5x_1 + 20x_2 + 8x_3$ subject to

$$\begin{array}{rcrcrcr} 3x_1 & + & x_2 & + & x_3 & \leq & 10 \\ 2x_1 & + & 3x_2 & + & 4x_3 & \geq & 20 \\ x_1 & & & - & x_3 & \leq & 0 \\ x_1 & \geq & 0, \ x_2 & & \geq 0, & x_3 & \geq 0. \end{array}$$

a) State and solve the dual problem (D). Determine the solution to (P).
b) Propose two other methods for solving (P).

Exercise 4.7. Consider the following linear programming problem where a and b denote two real parameters:
Maximize $f(p_1, p_2) = ap_1 - bp_2$ subject to

$$\begin{array}{rcrcr} p_1 & - & bp_2 & \leq & -1 \\ p_1 & - & p_2 & \leq & 1 \\ p_1 & \geq & 0, \ p_2 & \geq & 0. \end{array}$$

a) According to the value of the parameters, discuss the existence and the number of extreme points of the feasible set of (P).
b) Determine the extreme directions.
c) Using the necessary and sufficient conditions of optimality, discuss the

existence and the number of optimal solutions to (P), according to the value of the parameters.
d) State the dual problem (D).
e) Using the duality theorem, discuss the existence and the number of optimal solutions to (D), according to the value of the parameters.

Exercise 4.8. Consider the problem (P):
Maximize $f(x_1, x_2, x_3) = 8x_1 + 10x_2 + 13x_3$ subject to

$$\begin{array}{rcl} x_1 + x_2 + 5x_3 & \leq & 1 \\ 2x_1 + 3x_2 + x_3 & \leq & 1 \\ x_1 \geq 0, \; x_2 \geq 0, \; x_3 & \geq & 0. \end{array}$$

State the dual problem (D) of (P). Verify that $\widehat{x} = (4/9, 0, 1/9)$ is an optimal solution to (P) and determine an optimal solution to (D).

Exercise 4.9. Consider the problem (P):
Maximize $f(x_1, x_2, x_3) = 3x_1 + x_2 + 4x_3$ subject to

$$\begin{array}{rcl} 6x_1 + 3x_2 + 5x_3 & \leq & 25 \\ 3x_1 + 4x_2 + 5x_3 & \leq & 20 \\ x_1 \geq 0, \; x_2 \geq 0, \; x_3 & \geq & 0. \end{array}$$

a) Verify that $\widehat{x} = (5/3, 0, 3)$ is an optimal solution to (P).
b) State the dual problem (D).
c) Using the complementarity relations, determine an optimal solution $\widehat{p}$ to (D). Verify by geometric inspection.
d) Assume that the original problem is replaced by the problem (P'):
Maximize $f(x_1, x_2, x_3) = 3x_1 + 3x_2 + 4x_3$ subject to

$$\begin{array}{rcl} 6x_1 + 2x_2 + 5x_3 & \leq & 25 \\ 3x_1 + 3x_2 + 5x_3 & \leq & 20 \\ x_1 \geq 0, \; x_2 \geq 0, \; x_3 & \geq & 0. \end{array}$$

Is $\widehat{x}$ an optimal solutions to (P')?
e) Solve the dual problem (D') to (P').
f) Use the complementarity relations to determine the optimal solutions to (P').
g) Consider the problem (P''):
Maximize $f(x_1, x_2, x_3, u) = 3x_1 + x_2 + 4x_3 + 2u$ subject to

$$\begin{array}{rcl} 6x_1 + 3x_2 + 5x_3 + 3u & \leq & 25 \\ 3x_1 + 4x_2 + 5x_3 + 2u & \leq & 20 \\ x_1 \geq 0, \; x_2 \geq 0, \; x_3 \geq 0, \; u & \geq & 0. \end{array}$$

Is $(5/3, 0, 3, 0)$ an optimal solution to (P'')?
h) Let λ be a real parameter. Consider now the problem (P_λ):
Maximize $f_\lambda(x_1, x_2, x_3) = (3 + \lambda)x_1 + (1 + 4\lambda)x_2 + (4 + 2\lambda)x_3$ subject to

$$\begin{array}{rcl} 6x_1 + 3x_2 + 5x_3 & \leq & 25 \\ 3x_1 + 4x_2 + 5x_3 & \leq & 20 \\ x_1 \geq 0, \; x_2 \geq 0, \; x_3 & \geq & 0. \end{array}$$

Determine a real interval $[\underline{\lambda}, \overline{\lambda}]$ so that for every λ in this interval, $\widehat{x} = (5/3, 0, 3)$ be an optimal solution to Problem (P_λ).

i) Using the dual solution $\widehat{p}$ of $\widehat{x}$, find again the same condition.
j) Compute the minimum value of $1/(3x_1+x_2+4x_3)$ subject to the constraints of Problem (P).

Exercise 4.10. Use the simplex method for solving the following problem:
Minimize $f(x_1, x_2, x_3) = -2x_1 - x_2 - x_3$ subject to

$$\begin{array}{rcrcrcr} -4x_1 & - & 6x_2 & - & 3x_3 & \geq & -8 \\ -x_1 & + & 9x_2 & - & x_3 & \geq & 3 \\ 2x_1 & + & 3x_2 & - & 5x_3 & \geq & 4 \end{array}$$
$$x_1 \geq 0,\ x_2 \geq 0,\ x_3 \geq 0.$$

Exercise 4.11. Use the simplex method for solving the following problem:
Maximize $f(x_1, x_2, x_3) = 3x_1 + 2x_2 + 4x_3$ subject to

$$\begin{array}{rcrcrcr} x_1 & + & x_2 & + & 2x_3 & \leq & 4 \\ 2x_1 & & & + & 3x_3 & \leq & 5 \\ 2x_1 & + & x_2 & + & 3x_3 & \geq & 7 \end{array}$$
$$x_1 \geq 0,\ x_2 \geq 0,\ x_3 \geq 0.$$

State and solve the dual problem.

Exercise 4.12. Consider the problem:
Maximize $f(x_1, x_2, x_3) = x_1 + 3x_2 - x_3$ subject to

$$\begin{array}{rcrcrcr} 2x_1 & + & 2x_2 & - & x_3 & \leq & 10 \\ 3x_1 & - & 2x_2 & + & x_3 & \leq & 10 \\ x_1 & - & 3x_2 & + & x_3 & \leq & 10 \end{array}$$
$$x_1 \geq 0,\ x_2 \geq 0,\ x_3 \geq 0.$$

a) Determine the extreme points of the feasible set.
b) Show that the half-line defined by
$$x_1 = 0,\ x_2 = 5 + (1/2)t,\ x_3 = t,\ t \geq 0$$
is an extreme ray of the feasible set.
c) Show that the problem has no optimal solution.
d) Set the dual problem.
e) Verify the duality theorem.
f) Use the simplex method for solving the problem and its dual.

Exercise 4.13. Let A be a nonzero real $(m \times n)$-matrix, B be a real $(m \times p)$-matrix and C be a real $(m \times r)$-matrix. Set the dual problem of the following problem:
Find $x \in \mathbb{R}^n$, $y \in \mathbb{R}^p$, $z \in \mathbb{R}^r$ which maximize $\sum_{i=1}^n x_i$ subject to
$$Ax + By + Cz = 0$$
$$x \geq 0,\ y \geq 0.$$
Deduce that one and only one of the two following systems has at least one solution:
(I) $Ax + By + Cz = 0,\ x \geq 0,\ y \geq 0,\ z \neq 0$
(II) ${}^t uA > 0,\ {}^t uB \geq 0,\ {}^t uC = 0,\ u \in \mathbb{R}^m$.

Exercise 4.14 (Bondareva-Shapley's theorem). Let $N = \{1, 2, \ldots, n\}$ be a set of players, $\mathcal{N} = 2^N \setminus \{\emptyset\}$ be the collection of nonempty subsets of

N, $v : \mathcal{N} \to \mathbb{R}$ be a real function. An element $S \in \mathcal{N}$ is called *coalition* and $v(S)$ is interpreted as the total amount of transferable utility obtained by the coalition S. The pair (N, v) is called *cooperative game with transferable utility.* The *core*, $\mathcal{C}(N, v)$, of the game (N, v) is defined as the set of vectors $x \in \mathbb{R}^N$ such that

$$\sum_{i \in N} x_i = v(N) \text{ and } \sum_{i \in S} x_i \geq v(S), \ \forall S \in \mathcal{N}.$$

On the other hand, the cooperative game (N, v) is said to be *balanced* if $\forall \lambda = (\lambda_S)_{S \in \mathcal{N}} \in \mathbb{R}_+^{\mathcal{N}}$, the relations

$$\sum_{\{S \in \mathcal{N} | i \in S\}} \lambda_S = 1, \ \forall i \in N$$

imply

$$\sum_{S \in \mathcal{N}} \lambda_S v(S) \leq v(N).$$

Applying the duality theorem to the problem:

$$\text{Minimize } \sum_{i \in N} x_i \text{ subject to}$$

$$x \in \mathbb{R}^N, \ \sum_{i \in S} x_i \geq v(S), \ \forall S \in \mathcal{N}$$

prove that $\mathcal{C}(N, v)$ is nonempty if and only if the cooperative game (N, v) is balanced.

5. Convex Functions

This chapter is devoted to a class of functions from $\mathbb{R}^n$ into $\mathbb{R} \cup \{+\infty\}$ called convex functions and to give a first important property of such functions. Any convex function is continuous on the interior of its domain if this one is nonempty. If the domain of a convex function f has an empty interior, then the restriction of f to the affine set spanned by its domain is continuous on the relative interior of its domain (this expression makes sense because the domain of a convex function is a convex set).
Another nice property of convex functions is that for any pointwise converging sequence of real-valued convex functions defined on a relatively open set U, the limit function of this sequence is the uniform limit on any compact subset of U of this sequence.
Our definition of convex functions is less general than the one given by Rockafellar [51]. Rockafellar defines convexity for functions from $\mathbb{R}^n \to \overline{\mathbb{R}}$, by the convexity of their epigraph. But he shows and we will verify (Exercise 5.19) that for a function $\mathbb{R}^n \to \mathbb{R} \cup \{+\infty\}$, his definition of convexity coincides with the definition we will give below.

5.1 Basic definitions and properties

Let $f : A \to \mathbb{R} \cup \{+\infty\}$ be a function defined from A, a nonempty set of $\mathbb{R}^n$, into $\mathbb{R} \cup \{+\infty\}$. We require that A is convex. We say that f is *convex* on A if:

$$\forall a_1 \in A, \forall a_2 \in A, \forall \lambda \in]0,1[\,, f(\lambda a_1 + (1-\lambda)a_2) \leq \lambda f(a_1) + (1-\lambda) f(a_2).$$

If f is *real-valued* then f is convex on A if:

$$\forall a_1 \in A, \forall a_2 \in A, \forall \lambda \in [0,1]\,, f(\lambda a_1 + (1-\lambda)a_2) \leq \lambda f(a_1) + (1-\lambda) f(a_2).$$

The function f is said to be *strictly convex* if

$$\forall a_1 \in A, \forall a_2 \in A, \forall \lambda \in]0,1[\,, f(\lambda a_1 + (1-\lambda)a_2) < \lambda f(a_1) + (1-\lambda) f(a_2).$$

The *domain* of f is $\mathrm{dom}(f) := \{x \in \mathbb{R}^n \mid f(x) < +\infty\}$. Observe that $\mathrm{dom}(f)$ is *convex*.

The function $f : A \to \mathbb{R} \cup \{-\infty\}$ is *concave* on A if $-f$ is convex. Explicitly, f is concave on A if

$$\forall a_1 \in A, \forall a_2 \in A, \forall \lambda \in \,]0,1[\,, f(\lambda a_1 + (1-\lambda)a_2) \geq \lambda f(a_1) + (1-\lambda) f(a_2).$$

If f is *real-valued* then f is concave on A if:

$$\forall a_1 \in A, \forall a_2 \in A, \forall \lambda \in [0,1]\,, f(\lambda a_1 + (1-\lambda)a_2) \geq \lambda f(a_1) + (1-\lambda) f(a_2).$$

It is *strictly concave* on A if:

$$\forall a_1 \in A, \forall a_2 \in A, \forall \lambda \in \,]0,1[\,, f(\lambda a_1 + (1-\lambda)a_2) > \lambda f(a_1) + (1-\lambda) f(a_2).$$

Its *domain* will be : $\text{dom}(f) = \{x \in \mathbb{R}^n \mid f(x) > -\infty\}$.
We define the *epigraph* of a convex function $f : A \to \mathbb{R} \cup \{+\infty\}$:

$$\text{epi}(f) := \{(x, \mu) \in A \times \mathbb{R} \mid \mu \geq f(x)\}.$$

One can show that $f : A \to \mathbb{R} \cup \{+\infty\}$ is convex if, and only if, *epi*(f) is convex.
We have also the property that if $f : A \to \mathbb{R} \cup \{+\infty\}$ is convex then, $\forall \alpha \in \mathbb{R}$, the *level set* $\text{lev}_{\leq\alpha}(f) := \{x \in A \mid f(x) \leq \alpha\}$ is convex (see Exercise 5.6).
A convex function $f : A \to \mathbb{R} \cup \{+\infty\}$ can always be extended to a convex function on all of $\mathbb{R}^n$ by setting $f(x) = +\infty$ for $x \notin \text{dom}(f)$. Consequently, for simplicity, we will often consider that f is a convex function on $\mathbb{R}^n$.
For a concave function f, its *hypograph* is:

$$\text{hypo}(f) := \{(x, \mu) \in \mathbb{R}^n \times \mathbb{R} \mid \mu \leq f(x)\}.$$

As before, a function $f : A \to \mathbb{R} \cup \{-\infty\}$ is concave if, and only if, its hypograph is convex, and if $f : A \to \mathbb{R} \cup \{-\infty\}$ is concave then $\forall \alpha \in \mathbb{R}$, the *level set* $\text{lev}_{\geq\alpha}(f) := \{x \in A \mid f(x) \geq \alpha\}$ is convex.
Similarly, a concave function $f : A \to \mathbb{R} \cup \{-\infty\}$ can be extended to a concave function on all of $\mathbb{R}^n$ by setting $f(x) = -\infty$ for $x \notin \text{dom}(f)$.

We observe that since a convex function f on $\mathbb{R}^n$ takes an infinite value outside of its domain, the multiplication of f by a nonnegative scalar leads to the problem of arithmetic calculation of $0 \cdot (+\infty)$. As to the meaning of the function (λf) when $\lambda = 0$, we adopt the following rule:
(i) if $\lambda > 0$ then the function (λf) is defined by $(\lambda f)(x) = \lambda f(x), \forall x \in \text{dom}(f)$ and $(\lambda f)(x) = +\infty$ if $x \notin \text{dom}(f)$,
(ii) if $\lambda = 0$ then $(\lambda f) = 0$, i.e. $(\lambda f)(x) = 0, \forall x \in \mathbb{R}^n$.
With this rule, on can easily check that a function f is convex on $\mathbb{R}^n$ if, and only if

$$\forall x \in \mathbb{R}^n, \forall y \in \mathbb{R}^n, \forall \lambda \in [0,1], f(\lambda x + (1-\lambda)y) \leq (\lambda f)(x) + ((1-\lambda)f)(y).$$

Observe that we do not impose in this inequality λ to be in the interior of the interval $[0,1]$, but we use the notation $(\lambda f)(x)$ instead of $\lambda f(x)$. In this chapter and the next chapters 6 and 7, we will adopt this rule.

Proposition 5.1.1 (Jensen inequality). *Let U be a nonempty convex set of $\mathbb{R}^n$ and let $f : U \to \mathbb{R}\cup\{+\infty\}$ be a convex function on U. Then f is convex if, and only if, for any integer $p \geq 2$, for any p elements of U, $x_1, \ldots, x_p$, for any p positive numbers $\lambda_1, \ldots, \lambda_p$, the sum of which equals 1, then*

$$f(\lambda_1 x_1 + \ldots + \lambda_p x_p) \leq \lambda_1 f(x_1) + \ldots + \lambda_p f(x_p).$$

Proof. Let $f : U \to \mathbb{R} \cup \{+\infty\}$ where U is a convex, nonempty subset of $\mathbb{R}^n$. Assume f convex. Let $x_1, \ldots, x_p$ be p elements of U. If $p = 2$, then obviously $f(\lambda_1 x_1 + \ldots + \lambda_p x_p) \leq \lambda_1 f(x_1) + \ldots + \lambda_p f(x_p)$.
Assume that Jensen Inequality holds for $p - 1$. We will show that it holds also for p. If $\lambda_p = 1$, then the result is trivially true. So, we assume $\lambda_p \neq 1$. Let $s_{p-1} = \lambda_1 + \ldots + \lambda_{p-1}$. We have:

$$\lambda_1 x_1 + \ldots + \lambda_p x_p = s_{p-1}\left(\frac{\lambda_1}{s_{p-1}} x_1 + \ldots + \frac{\lambda_{p-1}}{s_{p-1}} x_{p-1}\right) + (1 - s_{p-1}) x_p.$$

The function f being convex, we have:

$$f(\lambda_1 x_1 + \ldots + \lambda_p x_p) \leq s_{p-1} f\left(\frac{\lambda_1}{s_{p-1}} x_1 + \ldots + \frac{\lambda_{p-1}}{s_{p-1}} x_{p-1}\right) + (1 - s_{p-1}) f(x_p).$$

We have assumed that Jensen Inequality holds for $p - 1$. Hence

$$f\left(\frac{\lambda_1}{s_{p-1}} x_1 + \ldots + \frac{\lambda_{p-1}}{s_{p-1}} x_{p-1}\right) \leq \frac{\lambda_1}{s_{p-1}} f(x_1) + \ldots + \frac{\lambda_{p-1}}{s_{p-1}} f(x_{p-1}).$$

Thus:

$$f(\lambda_1 x_1 + \ldots + \lambda_p x_p) \leq \lambda_1 f(x_1) + \ldots + \lambda_p f(x_p).$$

The converse is true by taking $p = 2$. □

Proposition 5.1.2. *(i) Let f, g be two convex functions from a convex set U, into $\mathbb{R} \cup \{+\infty\}$. Then $\max(f, g)$ is convex. More generally, consider a collection of convex functions $\{f_i\}_{i \in I}$, from U into $\mathbb{R}\cup\{+\infty\}$. Then $\sup\{f_i \mid i \in I\}$ is convex from U into $\mathbb{R} \cup \{+\infty\}$.*
(ii) If $(f_i)_{i \in \mathbb{N}}$ is a sequence of convex functions from $\mathbb{R}^n$ into $\mathbb{R}$ which converges pointwise, i.e. $\forall x \in \mathbb{R}^n$, the sequence $(f_i(x))_{i \in \mathbb{N}}$ converges in $\mathbb{R}$, then the function defined for all $x \in \mathbb{R}^n$ by $f(x) = \lim_{i \to +\infty} f_i(x)$ is convex from $\mathbb{R}^n$ into $\mathbb{R}$.
(iii) If $f : U \to \mathbb{R}$, $g : V \to \mathbb{R}$, with $f(U) \subset V$, are convex, if g is nondecreasing, then $g \circ f$ is convex.
(iv) If f, g are convex from a convex set U, into $\mathbb{R} \cup \{+\infty\}$ and if λ is a nonnegative real number, then $f + g$ and (λf) are convex.

Proof. See Exercise 5.1 □

5.2 Continuity theorems

In this section, we want to prove that for any convex function f from $\mathbb{R}^n$ into $\mathbb{R} \cup \{+\infty\}$, its restriction to $\text{aff}(\text{dom}(f))$ is continuous on $\text{ri}(\text{dom}(f))$. But f is *not* continuous on $\text{ri}(\text{dom}(f))$ if $\text{aff}(\text{dom}(f)) \neq \mathbb{R}^n$, i.e. if $\text{i}(\text{dom}(f)) = \emptyset$. Actually, for a convex function we have a stronger continuity result. We will prove that any convex function is locally Lipschitzian on the relative interior of its domain.
Recall that a *real-valued* function f defined on an open set U is *locally Lipschitzian* if for every $x \in U$, there exists an open ball $B(x, \varepsilon)$ and a constant $K(x)$ such that if $y, z \in B(x, \varepsilon)$ then $|f(y) - f(z)| \leq K(x) \|y - z\|$. If this property holds for any element in $V \subset U$ with a constant K independent of x,we say that f is *Lipschitzian* on V.
For a convex function f on $\mathbb{R}^n$, we say that it is locally Lipschitzian on the relative interior of its domain if for every $x \in \text{ri}(\text{dom}(f))$, there exists an open ball $B(x, \varepsilon)$ and a constant $K(x)$ such that if $y, z \in B(x, \varepsilon) \cap \text{aff}(\text{dom}(f))$ then $|f(y) - f(z)| \leq K(x) \|y - z\|$.
The proof of the continuity of the restriction of a convex function to the affine space spanned by its domain will be made in several steps. We first prove an interesting result (which also holds for the infinite dimension case, see Aubin [3]) that a real-valued convex function on an open set U is continuous if and only if it is bounded *above* on an open ball included in U. Secondly, we prove that any real-valued convex function on an open convex set is locally Lipschitzian and hence continuous. Hence if the interior of the domain of a convex function f is not empty, then f is continuous on the interior of its domain. Finally, when the interior of the domain of f is empty, then the restriction of f to the affine space spanned by its domain is continuous on the relative interior of its domain.
For simplicity, one can assume that $0 \in \text{ri}(\text{dom}(f))$. Indeed, let $g(x) = f(x + x_0)$ where $x_0 \in \text{ri}(\text{dom}(f))$ and $x \in \text{dom}(f) - \{x_0\}$. It is easy to check that g is convex, that $\text{aff}(\text{dom}(g)) = \text{aff}(\text{dom}(f)) - \{x_0\}$, that f is bounded on a relative neighborhood of x_0 if, and only if, g is bounded on a relative neighborhood of 0, and that f is continuous at x_0 if, and only if, g is continuous at 0. This trick will be used several times in this chapter. We will also assume that $\text{dom}(f) \neq \emptyset$.

Proposition 5.2.1. *Let f be a convex function from an open convex set U of $\mathbb{R}^n$ into $\mathbb{R}$. If f is bounded above on an open ball $B(x_0, \varepsilon)$, then f is locally bounded, (i.e., for every x in U, there exists an open ball centered at x with radius $\rho(x)$ on which f is bounded). Moreover, f is locally Lipschitzian on U.*

Proof. We first prove that if f is bounded above on an open ball $B(x_0, \varepsilon)$ then f is also bounded below. For simplicity, we assume that $x_0 = 0$, and that $\forall y \in B(0, \varepsilon)$, $f(y) \leq B$. Let $z \in B(0, \varepsilon)$. One has: $-z \in B(0, \varepsilon)$. Since

$0 = (1/2)z + (1/2)(-z)$, we have $f(0) \leq (1/2)f(z) + (1/2)f(-z)$. Therefore: $f(z) \geq 2f(0) - f(-z)$. But $\|-z\| < \varepsilon$, hence $f(-z) \leq B$ and $f(z) \geq 2f(0) - B$, which proves that f is bounded below on $B(0, \varepsilon)$.

Let $x \neq 0$. We now prove that f is bounded on an open ball centered in x. Choose $\rho > 1$ such that $z = \rho x \in U$. Let $\lambda = 1/\rho$. Then

$$B(x, (1-\lambda)\varepsilon) = \{v \in \mathbb{R}^n \mid v = (1-\lambda)y + \lambda z,\ y \in B(0, \varepsilon)\} \qquad \text{(figure 5.1).}$$

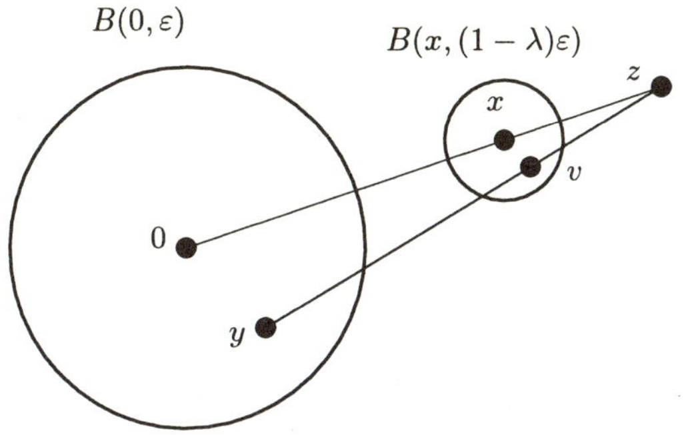

Fig. 5.1.

Moreover, for any $v \in B(x, (1-\lambda)\varepsilon)$, one has $f(v) \leq (1-\lambda)f(y) + \lambda f(z) \leq B + f(z)$. The function f is bounded above on $B(x, (1-\lambda)\varepsilon)$, is therefore bounded below on $B(x, (1-\lambda)\varepsilon)$ from the previous part of the proof. In other words, f is locally bounded: there exists M such that $|f|$ is bounded by M on the open ball $B(x, \varepsilon)$.

Finally, we show that f is Lipschitzian on $B(x, \varepsilon/2)$. Otherwise there exists $x_1, x_2 \in B(x, \varepsilon)$ such that

$$\frac{f(x_2) - f(x_1)}{\|x_2 - x_1\|} > \frac{4M}{\varepsilon}.$$

Define: $\alpha = \varepsilon/(2\|x_2 - x_1\|)$ and $x_3 = x_2 + \alpha(x_2 - x_1)$. One has: $\|x_3 - x_2\| = \varepsilon/2$ and $x_2 = (1/(1+\alpha))x_3 + (\alpha/(1+\alpha))x_1$. Hence: $f(x_2) \leq (1/(1+\alpha))f(x_3) + (\alpha/(1+\alpha))f(x_1)$. Since $\alpha = \|x_3 - x_2\|/\|x_2 - x_1\|$ in the previous inequality. We obtain that

$$\frac{f(x_3) - f(x_2)}{\|x_3 - x_2\|} \geq \frac{f(x_2) - f(x_1)}{\|x_2 - x_1\|} > \frac{4M}{\varepsilon}.$$

Since $\|x_3 - x_2\| = \varepsilon/2$, one has $f(x_3) - f(x_2) > 2M$, which contradicts $f(x_3) - f(x_2) \leq 2M$. □

Corollary 5.2.1. *Let f be a convex function from an open set U of $\mathbb{R}^n$ into $\mathbb{R}$. Then f is continuous on U if and only if f is bounded above on a ball $B(x, \varepsilon)$ included in U.*

Proof. From the previous proposition, if f is bounded above on a ball $B(x, \varepsilon)$ of U then f is locally Lipschitzian, and hence continuous. Conversely, if f is continuous on U, then obviously, for any $x \in U$, there exists an open ball $B(x, \varepsilon)$ included in U on which f is bounded above. □

Theorem 5.2.1. *Let f be a convex function from an open set U of $\mathbb{R}^n$ into $\mathbb{R}$. Then f is locally Lipschitzian on U and hence is continuous on U.*

Proof. Assume that $0 \in U$. One can choose $\alpha > 0$ sufficiently small such that the convex hull $V = \text{co}\{0, \alpha e^1, \dots, \alpha e^n\}$, where the e^i are the vectors of the canonical basis of $\mathbb{R}^n$, is contained in U. Recall that V has a nonempty interior (see Lemma 1.2.1) and thus contains an open ball B. Any x in V may be expressed as: $x = \lambda_0 0 + \alpha \sum_{i=1}^n \lambda_i e^i$ with $\lambda_i \geq 0, \forall i$, $\sum_{i=0}^n \lambda_i = 1$. Since f is convex,

$$f(x) \leq \lambda_0 f(0) + \sum_{i=1}^{n} \lambda_i f(\alpha e^i) \leq \max\left\{f(0), f(\alpha e^1), \dots, f(\alpha e^n)\right\}.$$

Therefore f is bounded above on B. The conclusion follows from Proposition 5.2.1. □

The following results are consequences of the previous proposition.

Corollary 5.2.2. *Let f be a convex function from $\mathbb{R}^n$ into $\mathbb{R}$. Then f is locally Lipschitzian and hence continuous.*

Proof. The domain of f is $\mathbb{R}^n$ which is open. Hence f is locally Lipschitzian and continuous. □

Corollary 5.2.3. *Let f be a convex function on $\mathbb{R}^n$. Then the restriction of f to* $\text{aff}(\text{dom}(f))$ *is locally Lipschitzian on* $\text{ri}(\text{dom}(f))$ *and hence continuous on* $\text{ri}(\text{dom}(f))$. *In particular, if* $\text{i}(\text{dom}(f)) \neq \emptyset$, *then f is locally Lipschitzian on* $\text{i}(\text{dom}(f))$ *and hence continuous on* $\text{i}(\text{dom}(f))$.

Proof. It is obvious if $\text{i}(\text{dom}(f)) \neq \emptyset$. If $\text{i}(\text{dom}(f)) = \emptyset$, the restriction of f to $\text{aff}(\text{dom}(f))$ is convex and its domain has a nonempty interior in $\text{aff}(\text{dom}(f))$. □

It is important to point out that if $\text{aff}(\text{dom}(f)) \neq \mathbb{R}^n$ for a convex function f from $\mathbb{R}^n$ into $\mathbb{R} \cup \{+\infty\}$, then, while the restriction of f to $\text{aff}(\text{dom}(f))$ is continuous on $\text{ri}(\text{dom}(f))$, this function is *not* continuous on $\text{ri}(\text{dom}(f))$ because $f(x) = +\infty, \forall x \notin \text{aff}(\text{dom}(f))$.

Consider the following example. Let $f : \mathbb{R}^2 \to \mathbb{R} \cup \{+\infty\}$ be defined by

$$f((x_1, x_2)) = \begin{cases} 0 & \text{if } (x_1, x_2) \in [0,1] \times \{0\}; \\ +\infty & \text{otherwise.} \end{cases}$$

Obviously, f is convex. We have $\text{dom}(f) = [0,1] \times \{0\}$, $\text{aff}(\text{dom}(f)) = \mathbb{R} \times \{0\}$ and $\text{ri}(\text{dom}(f)) =]0,1[\times \{0\}$. The restriction of f to $\text{aff}(\text{dom}(f))$ is continuous at $z_0 = (1/2, 0) \in \text{ri}(\text{dom}(f))$, but f is not continuous at z_0. Indeed, if the sequence $(z_n)_{n\in\mathbb{N}} = (1/2, 1/(n+1))_{n\in\mathbb{N}}$ converges to z_0 when n tends to $+\infty$, we have $f(z_n) = +\infty, \forall n \in \mathbb{N}$ while $f(z_0) = 0$.

5.3 Continuity properties of collections of convex functions

We now turn to the continuity properties of collections of real-valued convex functions. We want to show two results.
The first one is that for any function f defined on $U \times T$ where U is a relatively open convex nonempty subset of $\mathbb{R}^n$, T is a compact set of $\mathbb{R}^m$, if f is convex on U and continuous on T, then f is actually jointly continuous on $U \times T$.
The second result is that for any sequence of real-valued convex functions on a relatively open convex set U which converges pointwise, the limit function is convex and is the uniform limit of this sequence on any compact subset of U.

We now give some definitions.
Let U be a nonempty subset of $\mathbb{R}^n$. We say that the collection $\{f_i\}_{i\in I}$ of real-valued functions defined on U is *pointwise-bounded* if $\forall x \in U, \exists K(x)$ such that $\forall i \in I, |f_i(x)| \leq K(x)$.
A collection $\{f_i\}_{i\in I}$ of real-valued functions defined on U is *equi-Lipschitzian* on U if there exists $\alpha > 0$ such that $\forall i \in I, \forall x \in U, \forall y \in U$, $|f_i(x) - f_i(y)| \leq \alpha \|y - x\|$.
The two next propositions are useful for the proof of the two results we mentioned before.

Proposition 5.3.1. *Let U be a relatively open convex set of $\mathbb{R}^n$ and let $\{f_i\}_{i\in I}$ be a collection of convex functions from U into $\mathbb{R}$, pointwise-bounded on U. Then $\{f_i\}_{i\in I}$ is uniformly bounded on any compact subset of U.*

Proof. Without loss of generality, one can assume U open. Take some compact subset K of U. Let $f(x) = \sup\{f_i(x) \mid i \in I\}$. This function is convex (Proposition 5.1.2). Since the collection $\{f_i\}_{i\in I}$ is pointwise-bounded, f is real-valued, hence continuous on U (Theorem 5.2.1), and bounded from above on K. Thus $\{f_i\}_{i\in I}$ is uniformly bounded from above on K.
To prove that $\{f_i\}_{i\in I}$ is uniformly bounded from below on K, it suffices to show that the collection $\{f_i\}_{i\in I}$ is bounded from below on U by a continuous function g.
Take some $x_0 \in U$. Let $\beta = \inf\{f_i(x_0) \mid i \in I\}$. Choose $\varepsilon > 0$ small

enough such that $\{x_0\} + \varepsilon\overline{B(0,1)} \subset U$, where $\overline{B(0,1)}$ is the unit-ball of $\mathbb{R}^n$. Let $\alpha = \max\{|f(x)| \mid x \in \{x_0\} + \varepsilon\overline{B(0,1)}\}$. Given $x \in U, x \neq x_0$, define $\lambda = \varepsilon/(\varepsilon + \|x_0 - x\|)$, $z = x_0 + (\varepsilon/\|x_0 - x\|)(x_0 - x)$. Then we have $x_0 = \lambda x + (1-\lambda)z$. Since $\|z - x_0\| = \varepsilon$, $0 < \lambda < 1$, we have

$$\forall i, \beta \leq f_i(x_0) \leq (1-\lambda)f_i(z) + \lambda f_i(x) \leq (1-\lambda)\alpha + \lambda f_i(x) \leq \alpha + \lambda f_i(x).$$

Thus

$$\forall i \in I,\ f_i(x) \geq (\beta - \alpha)/\lambda = (\beta - \alpha)(\varepsilon + \|x_0 - x\|)/\varepsilon.$$

The function g defined by $\forall x \in U, g(x) = (\beta-\alpha)(\varepsilon+\|x_0-x\|)/\varepsilon$ is continuous. The proof is complete. □

Proposition 5.3.2. *In the conditions of Proposition 5.3.1, the collection $\{f_i\}_{i\in I}$ is equi-Lipschitzian on any compact subset of U.*

Proof. As before, we assume, for simplicity, that U is open. Let K be a compact subset of U. There exists $\varepsilon > 0$ sufficiently small such that $K + \varepsilon\overline{B(0,1)}$ is included in U, where $\overline{B(0,1)}$ is the unit-ball. From the previous proposition, there exist α, β such that

$$\forall x \in K + \varepsilon\overline{B(0,1)}, \forall i \in I,\ \beta \leq f_i(x) \leq \alpha.$$

Let $x \in K, y \in K, y \neq x$, and $z = y + (\varepsilon/\|y-x\|)(y-x)$. Then $\|z-y\| = \varepsilon$ and hence $z \in K + \varepsilon\overline{B(0,1)}$. Since $y = (1-\lambda)x + \lambda z$, where $\lambda = \|y-x\|/(\varepsilon+\|y-x\|)$, we have

$$\forall i \in I,\ f_i(y) \leq (1-\lambda)f_i(x) + \lambda f_i(z) = f_i(x) + \lambda(f_i(z) - f_i(x)),$$

and consequently

$$\forall i \in I,\ f_i(y) - f_i(x) \leq \lambda(f_i(z) - f_i(x)) \leq \lambda(\alpha - \beta) \leq (\alpha - \beta)(\|y - x\|/\varepsilon).$$

Since x, y are arbitrarily chosen in K, we have actually

$$\forall i \in I, \forall x \in K, \forall y \in K,\ \|f_i(y) - f_i(x)\| \leq (\alpha - \beta)(\|y - x\|/\varepsilon).$$ □

Theorem 5.3.1. *Let U be a relatively open convex set of $\mathbb{R}^n$ and let T be a compact subset of $\mathbb{R}^m$. Let f be a real-valued function on $U \times T$ such that $f(.,t)$ is convex for each $t \in T$, $f(x,.)$ is continuous for each $x \in U$. Then f is continuous on $U \times T$.*

Proof. For each $x \in U$, the function $f(x,.)$ is continuous on T and hence bounded on T since T is compact. From the previous proposition, the collection $\{f(.,t)\}_{t\in T}$ is equi-Lipschitzian on any compact subset of U.
Let $(x_0, t_0) \in U \times T$. Take a compact subset K of U which contains x_0. We have

$$\exists \alpha > 0, \forall t \in T, \forall x \in K,\ |f(x,t) - f(x_0,t)| \leq \alpha\|x - x_0\|.$$

Hence, given $\varepsilon > 0$, there exists $\delta > 0$ such that $\forall t \in T$, if $x \in K$, $\|x - x_0\| \leq \delta$ then $|f(x,t) - f(x_0,t)| \leq \varepsilon/2$. Since the function $f(x_0,.)$ is continuous at t_0, there exists a neighborhood V of t_0 in T such that $|f(x_0,t) - f(x_0,t_0)| \leq \varepsilon/2$, $\forall t \in V$. Consequently, for any $(x,t) \in K \times V$ such that $\|x - x_0\| \leq \delta$, we have $|f(x,t) - f(x_0,t_0)| \leq |f(x,t) - f(x_0,t)| + |f(x_0,t) - f(x_0,t_0)| \leq \varepsilon/2 + \varepsilon/2 = \varepsilon$. □

Theorem 5.3.2. *Let U be a relatively open convex set of $\mathbb{R}^n$ and let $(f_i)_{i\in\mathbb{N}}$ be a sequence of real-valued convex functions on U. Assume that for every $x \in U$, $\lim_{i\to+\infty} f_i(x)$ exists in $\mathbb{R}$. Define for all x in U the function f by $f(x) = \lim_{i\to+\infty} f_i(x)$.*
(i) Then f is convex, real-valued on U and is the uniform limit on any compact subset of U of the sequence$(f_i)_{i\in\mathbb{N}}$.
(ii) Moreover, if $(x_i)_{i\in\mathbb{N}}$ is a sequence of points of U converging to a point $x \in U$, then $f_i(x_i)$ converges to $f(x)$.

Proof. It is easy to check that f is convex, real-valued on U.
We first remark that since $\forall x \in U$, the sequence $(f_i(x))_{i\in\mathbb{N}}$ is convergent, the collection $\{f_i\}_{i\in\mathbb{N}}$ is pointwise-bounded on U. Let K be a compact subset of U. From Proposition 5.3.2, the collection $\{f_i\}_{i\in\mathbb{N}}$ is equi-Lipschitzian on K, i.e.

$$\exists \alpha > 0, \forall i \in \mathbb{N}, \forall x \in K, \forall y \in K, \ |f_i(x) - f_i(y)| \leq \alpha\|y - x\|.$$

Therefore, given $\varepsilon > 0$, there exists $\delta > 0$ such that $\forall x \in K$, $\|y - x\| \leq \delta \Rightarrow \forall i \in \mathbb{N}$, $|f_i(x) - f_i(y)| \leq \varepsilon/3$. By a classical compactness argument, there exists $z_1, \ldots, z_p$ in K such that $K \subset B(z_1,\delta) \cup \ldots \cup B(z_p,\delta)$. Since the sequences $(f_i(z_k))_{i\in\mathbb{N}}$ converge, there exists N such that $\forall i \geq N, \forall j \geq N$, $|f_i(z_k) - f_j(z_k)| \leq \varepsilon/3$, $k = 1, \ldots, p$.

Consequently, for each $x \in K$, we can take some z_k which satisfies $\|z_k - x\| \leq \delta$. We have $|f_i(x) - f_j(x)| \leq |f_i(x) - f_i(z_k)| + |f_i(z_k) - f_j(z_k)| + |f_j(z_k) - f_j(x)| \leq \varepsilon/3 + \varepsilon/3 + \varepsilon/3 = \varepsilon$. Let j tends to $+\infty$. We have $|f_i(x) - f(x)| \leq \varepsilon$, $\forall x \in K, \forall i \geq N$. The proof of Assertion (i) is complete.

(ii) For simplicity, assume U open. Fix some i_0 in $\mathbb{N}$. Let K be a compact set of U which contains in its interior x and $x_i, \forall i \geq i_0$. We have just shown in (i) that the collection $\{f_i\}_{i\in\mathbb{N}}$ is pointwise-bounded. From Proposition 5.3.2, this collection is equi-Lipschitzian on K. Hence $\exists\alpha, \forall i \geq i_0$, $|f_i(x) - f_i(x_i)| \leq \alpha\|x - x_i\|$. Thus, given $\varepsilon > 0$, there exists i_1 such that $\forall i \geq i_1$, $|f_i(x) - f_i(x_i)| \leq \varepsilon/2$. Now, since the sequence $(f_i(x))_{i\in\mathbb{N}}$ converges to $f(x)$, there exists i_2 such that $\forall i \geq i_2$, $|f_i(x) - f(x)| \leq \varepsilon/2$. Consequently, $\forall i \geq \max\{i_1, i_2\}$, $|f_i(x_i) - f(x)| \leq \varepsilon$. □

Exercises

Exercise 5.1. Demonstrate Proposition 5.1.2

Exercise 5.2. Show that the function $f : \mathbb{R}^2 \to \mathbb{R}$ defined by $f(x_1, x_2) = x_1 x_2$ is not convex on $\mathbb{R}^2$.

Exercise 5.3. Show that $|x^p|$, where p is an integer ≥ 1, is convex on $\mathbb{R}$. Prove that $\sum_{i=1}^{n} |x_i^p|$ is convex on $\mathbb{R}^n$.

Exercise 5.4. Show that if f is convex from a convex set $U \subset \mathbb{R}^2$ into $\mathbb{R} \cup \{+\infty\}$, then $f(., x_2)$ is convex on $V = \{x_1 \mid (x_1, x_2) \in U\}$.
The converse may be false: $f(., x_2)$ and $f(x_1, .)$ may be convex while f is not convex with respect to (x_1, x_2).

Exercise 5.5. Let U be a nonempty convex set of $\mathbb{R}^n$. Show that the following functions are convex on $\mathbb{R}^n$:
$\delta_U^*(x) = \sup\{x \cdot y \mid y \in U\}$ *(Support function)*;
$\gamma_U(x) = \inf\{\lambda \geq 0 \mid x \in \lambda U\}$ *(Gauge)*;
$d_U(x) = \inf\{\|x - y\| \mid y \in U\}$ *(Distance)*.

Exercise 5.6. Recall the definition of the *epigraph* of a convex function $f : U \to \mathbb{R}$, where U is a convex set of $\mathbb{R}^n$:

$$\mathrm{epi}(f) = \{(x, \mu) \in U \times \mathbb{R} \mid \mu \geq f(x)\}.$$

a) Show that f is convex if, and only if, *epi*(f) is convex.
b) Show that if f is convex then, $\forall \alpha \in \mathbb{R}$, the level sets $\mathrm{lev}_{\leq \alpha}(f) = \{x \in \mathbb{R}^n \mid f(x) \leq \alpha\}$, $\mathrm{lev}_{< \alpha}(f) = \{x \in \mathbb{R}^n \mid f(x) < \alpha\}$ are convex.

Exercise 5.7. Let $f : U \to \mathbb{R}$ be a continuous function on a convex, nonempty set U of $\mathbb{R}^n$. Show that f is convex on U if, and only if, $\forall x \in U, \forall y \in U$, $f((x + y)/2) \leq (f(x) + f(y))/2$.

Exercise 5.8. A function $f : \mathbb{R}^n \to \mathbb{R} \cup \{+\infty\}$ is *positively homogenous* if $\forall x \in \mathbb{R}^n, \forall \lambda > 0$, $f(\lambda x) = \lambda f(x)$. An example of positively homogenous function which is not simply linear is $f(x) = |x|, \forall x \in \mathbb{R}$.
a) Show that a positively homogenous function f is convex if, and only if, $\forall x \in \mathbb{R}^n, \forall y \in \mathbb{R}^n$, $f(x + y) \leq f(x) + f(y)$.
b) Show also that if f is positively homogeneous convex function on $\mathbb{R}^n$ then $f(\lambda_1 x_1 + \ldots + \lambda_m x_m) \leq \lambda_1 f(x_1) + \ldots + \lambda_m f(x_m)$,
$\forall (x_1, \ldots, x_m) \in (\mathbb{R}^n)^m, \forall (\lambda_1, \ldots, \lambda_m)$ with $\lambda_i > 0, \forall i = 1, \ldots, m$.

Exercise 5.9. a) Show that the function ln is concave.
b) Show that, if p, q are two positive real numbers such that $1/p + 1/q = 1$ and if a, b are two strictly positive numbers, then one has:

$$a^{1/p} b^{1/q} \leq \frac{a}{p} + \frac{b}{q}.$$

c) Prove that, for all $x = (x_1, \ldots, x_n)$ and for all $y = (y_1, \ldots, y_n)$, one has:

$$\sum_{i=1}^{n} |x_i y_i| \leq \left(\sum_{i=1}^{n} |x_i|^p\right)^{1/p} \left(\sum_{i=1}^{n} |y_i|^q\right)^{1/q} \tag{5.1}$$

where p, q are two positive numbers verifying $1/p + 1/q = 1$.
Hint:
Define $a_i = |x_i|^p / (\sum_{j=1}^{n} |x_j|^p)$, $b_i = |y_i|^q / (\sum_{j=1}^{n} |y_j|^q)$ and apply b).
d) Prove that:

$$(\sum_{i=1}^{n} |x_i + y_i|^p)^{1/p} \leq (\sum_{i=1}^{n} |x_i|^p)^{1/p} + (\sum_{i=1}^{n} |y_i|^p)^{1/p}$$

Hint:
Observe that:

$$\sum_{i=1}^{n} |x_i + y_i|^p \leq \sum_{i=1}^{n} |x_i + y_i|^{(p-1)} |x_i| + \sum_{i=1}^{n} |x_i + y_i|^{(p-1)} |y_i|.$$

Apply (5.1) to each sum of the second member of the previous inequality with p and $q = p/(p-1)$.
e) Deduce from d) that the function:

$$x = (x_1, \ldots, x_n) \to (\sum_{i=1}^{n} |x_i|^p)^{1/p}$$

is convex.

Exercise 5.10. Let C be a convex set of $\mathbb{R}^n$ which contains the origin. Let γ_C denote the gauge as in Exercise 5.5. Let

$$K_C = \bigcup_{\lambda \geq 0} (\lambda C) \text{ and } K_\infty = \bigcap_{\lambda > 0} (\lambda C).$$

Show that:
$\gamma_C(x) = 0 \Leftrightarrow x \in K_\infty$ and $\gamma_C(x) < +\infty \Leftrightarrow x \in K_C$.
If C is bounded then: $\gamma_C(x) = 0 \Leftrightarrow x = 0$.

Exercise 5.11. Let f be a convex function from a convex set $U \subset \mathbb{R}^n$ into $\mathbb{R} \cup \{+\infty\}$. Let x, y be two elements in $\mathrm{i}(\mathrm{dom}(f))$.
Show that the function $g : t \in [0,1] \to f(tx + (1-t)y)$ is convex and continuous.

Exercise 5.12. If $f : \mathbb{R}^n \to \mathbb{R}$ is a convex function such that $\forall \lambda \in \mathbb{R}$, $S_\lambda = \{x \mid f(x) \leq \lambda\}$ is compact, then we say that f is *inf-compact*. Show that f is inf-compact if, and only if: $\mathrm{A}(epi(f)) \cap \{(u,t) \in \mathbb{R}^n \times \mathbb{R} \mid t = 0\} = \{0\}$, where $\mathrm{A}(epi(f))$ is the asymptotic cone of $(\mathrm{epi}(f))$.

Exercise 5.13. Let A be a convex set of $\mathbb{R}^n$ and let f be a function from A into $\mathbb{R} \cup \{+\infty\}$. Show that f is convex on A if, and only if, for all a and all a' in A, the function φ defined for all $t \in [0,1]$ by $\varphi(t) = f(ta + (1-t)a')$ is convex in $[0,1]$.

Exercise 5.14. a) Let f be a *convex, differentiable* function on the interval I of $\mathbb{R}$. Let $t < t'$ be two points in I. Let $x \in]t, t'[$. Show that:

$$\frac{f(x) - f(t)}{x - t} \leq \frac{f(t') - f(t)}{t' - t} \leq \frac{f(x) - f(t')}{x - t'}. \tag{5.2}$$

Let t_1, t_2 in I verify $t_1 < t_2$. Let x_1, x_2, two points in $]t_1, t_2[$, verify $x_1 < x_2$. Deduce from (5.2) that

$$\frac{f(x_1) - f(t_1)}{x_1 - t_1} \leq \frac{f(x_2) - f(t_1)}{x_2 - t_1} \leq \frac{f(t_1) - f(t_2)}{t_1 - t_2} \leq \frac{f(x_2) - f(t_2)}{x_2 - t_2}. \tag{5.3}$$

Show that $f'(t_1) \leq f'(t_2)$.
b) We now assume that f' is nondecreasing. Show that f is convex.
Hint: Take two points $t_1, t_2 \in I, t_1 < t_2$ and $\lambda \in [0,1]$. Use the Mean Value theorem (Theorem A.2.2 in Appendix) to prove that:

$$f(\lambda t_1 + (1-\lambda)t_2) - f(t_1) = (1-\lambda)(t_2 - t_1)f'(\mu_1)$$
$$f(t_2) - f(\lambda t_1 + (1-\lambda)t_2) = \lambda(t_2 - t_1)f'(\mu_2)$$

with $t_1 \leq \mu_1 \leq \lambda t_1 + (1-\lambda)t_2 \leq \mu_2 \leq t_2$.
Conclude that f is convex.
c) Assume that f is twice differentiable in I. Show that f is convex if, and only if, $f''(t) \geq 0$, $\forall t \in I$.
d) Let f be a function which is differentiable on an open convex set A of $\mathbb{R}^n$. Prove that it is convex on A if, and only if, for all a, a' in A, the function g defined for $t \in [0,1]$ by:

$$g(t) = \sum_{i=1}^{n} (a_i - a'_i) \frac{\partial f}{\partial x_i}(ta + (1-t)a')$$

is a nondecreasing function, where $(a_i), (a'_i)$ are the components of a and a'.
Hint: Consider the function $\varphi(t) = f(ta + (1-t)a')$ for $t \in [0,1]$.
e) Let f be a twice differentiable function on an open convex set A of $\mathbb{R}^n$. Show that f is convex on A if, and only if, for any $a_0 \in A$, the following quadratic form:

$$\Psi(h) = \sum_{i,j} h_i h_j \frac{\partial^2 f}{\partial x_i \partial x_j}(a_0)$$

is nonnegative, i.e., $\Psi(h) \geq 0$, $\forall h = (h_1, \ldots, h_n)$.
Hint: Take $a, a' \in A$. Consider the function $\varphi(t) = f(ta + (1-t)a')$ for $t \in [0,1]$.

Exercise 5.15. Let $f : \mathbb{R} \to \mathbb{R}$ be a convex function. a) Assume that $f(x) \to 0$ when $x \to +\infty$. Show that f is nonincreasing and $f(x) \geq 0$, $\forall x \in \mathbb{R}$.
b) Show that if $\lim_{x\to+\infty}(f(x) - \alpha x - \beta) = 0$ then $f(x) \geq \alpha x + \beta$, $\forall x \in \mathbb{R}$ (i.e. if the line $y = \alpha + \beta$ is an asymptote on the right, then the function is above its asymptote).

Exercise 5.16. Show that the following functions $f : \mathbb{R} \to \mathbb{R} \cup \{+\infty\}$ are convex:
a) $f(x) = e^{\alpha x}, \alpha \in \mathbb{R}$.
b) $f(x) = x^p$ if $x \geq 0, f(x) = +\infty$ if $x < 0$, where $p \geq 1$.
c) $f(x) = -x^p$ if $x \geq 0, f(x) = +\infty$ if $x < 0$, where $p \in [0, 1]$.
d) $f(x) = x^p$ if $x > 0, f(x) = +\infty$ if $x \leq 0$, where $p \leq 0$.
e) $f(x) = (\alpha^2 - x^2)^{-1/2}$ if $|x| < \alpha, f(x) = +\infty$ if $|x| \geq \alpha$, where $\alpha > 0$.
f) $f(x) = -\ln(x)$ if $x > 0, f(x) = +\infty$ if $x \leq 0$.

Exercise 5.17. Show that, if f is convex, real-valued and locally Lipschitzian on an open convex set $U \subset \mathbb{R}^n$, then f is Lipschitzian on any compact set of U.

Exercise 5.18. a) **Preliminaries.** Let U be a non-empty subset of $\mathbb{R}^n$, and f be a function from U into $\overline{\mathbb{R}}$. For every $x_0 \in \overline{U}$, we define:

$$\liminf_{x\to x_0} f(x) = \lim_{\varepsilon\to 0_+} (\inf\{f(x) \mid x \in B(x_0, \varepsilon) \cap U\}),$$

where $B(x_0, \varepsilon)$ is the closed ball centered in x_0 with radius ε.
One notices that if $x_0 \in U$ then $\liminf_{x\to x_0} f(x) \leq f(x_0)$.
(i) Show that the previous definition is meaningful.
(ii) One cannot replace lim inf by lim because the limit does not always exist as shown by the following example:
Let $U = \mathbb{R}^2_+$. Let f be defined by:

$$\forall (x, y) \in U,\ f(x, y) = \begin{cases} \dfrac{x^2 + y^2}{2y} & \text{if } (x, y) \neq (0, 0); \\ 1 & \text{if } (x, y) = (0, 0). \end{cases}$$

Compute $\liminf_{(x,y)\to(0,0)} f(x, y)$. Show that $\lim_{(x,y)\to(0,0)} f(x, y)$ does not exist.
(iii) A function f defined on U is said to be *lower semi-continuous (lsc)* if $\forall x \in U,\ f(x) = \liminf_{y\to x} f(y)$.
Show that f is lsc if, and only if, $\forall x,\ f(x) \leq \lim f(x_n)$ for any sequence $\{x_n\}$ converging to x such that the sequence $(f(x_n))_{n\in\mathbb{N}}$ has a limit.
(iv) Prove the following equivalence: f is lsc $\Leftrightarrow \forall \alpha \in \mathbb{R}$, $\{x \mid f(x) \leq \alpha\}$ is closed.
(v) A real-valued function f is said to be *upper semi-continuous (usc)* if $-f$ is lsc. Show that f is continuous if, and only if, it is lsc and usc.

b) We now assume that U is convex and that f is convex from U into $\mathbb{R}$.
(i) Let x_0 is a point of $\mathrm{Bd}^{\mathrm{r}}(U)$. Show that $\liminf_{x\to x_0} f(x) > -\infty$.

Let $x_0 \in \mathrm{ri}(U)$. Show that $\liminf_{x\to x_0} f(x) = f(x_0)$.
(ii) We assume furthermore that U is open. Let V be the set of points $y \in \overline{U}$ such that $\liminf_{x\to y} f(x) < +\infty$. Let $\tilde{f}$ be defined by: $\tilde{f}(y) = \liminf_{x\to y} f(x)$, $\forall y \in V$.
Show that V is convex and that $\tilde{f}$ is convex on V.

c) Let f be a lsc, convex function from $\mathbb{R}^n$ into $\mathbb{R} \cup \{+\infty\}$. Let $x \in \mathrm{dom}(f)$. Show that $\forall y \in \mathbb{R}^n$, $f(y) = \lim_{\lambda\to 1,\lambda<1} f(\lambda y + (1-\lambda)x)$.

Exercise 5.19. The purpose of this exercise is to show that the definition of convexity given by Rockafellar [51] for functions from $\mathbb{R}^n$ into $\overline{\mathbb{R}}$ coincides with the definition we give in this chapter when f takes values in $\mathbb{R} \cup \{+\infty\}$.
Let f be a function from $\mathbb{R}^n$ into $\overline{\mathbb{R}}$. We say that f is *convex in the sense of Rockafellar* if its epigraph is convex, where the epigraph of f is the set $\{(x,\mu) \in \mathbb{R}^n \times \mathbb{R} \mid \mu \geq f(x)\}$.
a) Show that f is convex in the sense of Rockafellar if, and only if

$$f((1-\lambda)x + \lambda y) < (1-\lambda)\alpha + \lambda\beta, 0 < \lambda < 1$$

whenever $\alpha > f(x), \beta > f(y)$.
b) Assume that f takes values in $\mathbb{R}\cup\{+\infty\}$. Show that epi$(f)$ is convex if, and only if $f((1-\lambda)x+\lambda y) \leq (1-\lambda)f(x)+\lambda f(y)$, $\forall x \in \mathbb{R}^n, \forall y \in \mathbb{R}^n, \forall \lambda \in]0,1[$.

Exercise 5.20. We first give some definitions.
A mapping $f : A \to B$ where A, B are nonempty subsets of $\mathbb{R}^n$, is an *homeomorphism* between A and B if f is continuous, one-to-one and if its inverse f^{-1} is continuous. Two nonempty subsets A and B of $\mathbb{R}^n$ are *homeomorphic* if there exists an homeomorphism between them.
In this exercise we want to prove that if C is a convex, compact subset of $\mathbb{R}^n$, with nonempty interior, then C and the closed unit-ball $\overline{B(0,1)}$ are homeomorphic.
Without loss of generality, we assume $0 \in \mathrm{i}(C)$ and $C \subset \overline{B(0,1)}$.
Consider the gauge function γ_C defined by $\forall x \in \mathbb{R}^n$, $\gamma_C(x) = \inf\{\lambda > 0 \mid x \in \lambda C\}$. We know that $\gamma_C(0) = 0, \gamma_C(x) > 0$, $\forall x \neq 0$, $\gamma_C(x) \leq 1$ if $x \in C$, and $\forall x \neq 0$, $(1/\gamma_C(x))x \in C$ (see Exercise 1.16). Since γ_C is convex, real-valued, it is continuous (Theorem 5.2.1).
Define the mapping $r : \mathbb{R}^n \to \mathbb{R}^n$ as follows: $r(x) = (\|x\|/\gamma_C(x))x$ if $x \neq 0$ and $r(0) = 0$.
a) Show that $\gamma_C(x) \geq 1$, $\forall x \in S(0,1)$ (the unit-sphere). Prove that r is continuous, and $r(x) \in C$ if $x \in \overline{B(0,1)}$.
b) Let $\rho : \mathbb{R}^n \to \mathbb{R}^n$ defined by $\rho(y) = (\gamma_C(y)/\|y\|)y$ for all $y \neq 0$ and $\rho(0) = 0$.
Show that ρ is continuous, $\rho(y) \in \overline{B(0,1)}$ if $y \in C$ and ρ is the inverse mapping of r. Conclude that $\overline{B(0,1)}$ and C are homeomorphic.
c) Show that moreover r is an homeomorphism between $S(0,1)$ an ∂C (the boundary of C).

6. Differential Theory of Convex Functions

In this chapter, we give three notions of differentiability for convex functions. The first one is the directional derivative which always exists in $\mathbb{R} \cup \{+\infty\}$ at any point in the relative interior of the domain of a convex function. While the differential of a function is a linear functional, the directional derivative is only a sublinear function.

The second notion is the subgradient. When a convex function f is finite at x, a subgradient x^* of f at x generates the function $\mathbb{R}^n \to \mathbb{R}$ defined by $\forall z \in \mathbb{R}^n$, $h(z) = f(x) + x^* \cdot (z - x)$, the graph of which is a supporting hyperplane to the epigraph of f at $(x, f(x))$. The set of subgradients of f at x is called subdifferential of f at x. There is a relationship between the directional derivative and the subdifferential. At any point x in the relative interior of a convex function f, the directional derivative of f at x is the support function of the subdifferential of f at x.

The third notion is well-known. That is the differentiability of a function. A convex function is differentiable at x if, and only if, x is in the interior of its domain, and the directional derivative is a linear functional or, equivalently, the subdifferential is a singleton.

Another interesting property of convex functions is that if a convex function is differentiable in an open set, then, it is actually continuously differentiable in this set. In Chapter 5, we have shown that any pointwise limit of a sequence of convex functions defined on the same open set U is convex, continuous, and is the uniform limit on any compact subset of U. In this chapter, we obtain more. For such a sequence of convex functions, their subdifferentials at x "converge" (the meaning of this word will be made precise in this chapter) to the subdifferential at x of the limit function. If the functions of this sequence and their limit function are differentiable at x, then their gradients at x converge (in the usual sense) to the gradient at x of the limit function.

To obtain these results, two main tools will be used. The first one is the Separation theorem (in its first or second form), the second one is the Hahn-Banach theorem. The Separation theorem has been presented in Chapter 2 of this book. We will prove the Hahn-Banach theorem in the following section. We emphasize that the Hahn-Banach theorem and the Separation theorem are in fact equivalent. We will prove the Hahn-Banach theorem by using the

Separation theorem. But one can deduce the Separation theorem from the Hahn-Banach theorem (see, e.g. Rudin [53]). We will prove (Exercise 6.17) that the separation theorems studied in Chapter 2, the Hahn-Banach theorem and the subdifferentiability theorem established in this chapter form a loop of equivalent results.

Finally, let us mention that generalizations of directional derivatives and of subgradients to nonconvex functions can be found in Clarke [14].

6.1 The Hahn-Banach dominated extension theorem

Theorem 6.1.1. *Let f be a convex function from $\mathbb{R}^n$ into $\mathbb{R} \cup \{+\infty\}$ such that $0 \in \mathrm{ri}(\mathrm{dom}(f))$. Let M be a vector subspace of $\mathbb{R}^n$. Let h be a linear functional on M such that:*

$$\forall x \in M \cap \mathrm{ri}(\mathrm{dom}(f)),\ h(x) \leq f(x).$$

Then there exists a linear extension h_1 of h on $\mathbb{R}^n$ which verifies:

$$\forall x \in \mathbb{R}^n,\ h_1(x) \leq f(x).$$

Proof. We will first prove the result for the case $0 \in \mathrm{i}(\mathrm{dom}(f))$. Let $U = \mathrm{i}(\mathrm{dom}(f))$. Define $A = \{(x,t) \in U \times \mathbb{R} \mid f(x) < t\}$ and $B = \{(x,t) \in M \times \mathbb{R} \mid h(x) = t\}$. The set B is a vector space and A is nonempty since for all $\varepsilon > 0$, one has $(0, f(0) + \varepsilon) \in A$. It is obvious that $A \cap B = \emptyset$. Hence, by the first Separation theorem (Proposition 2.1.7), there exists $(p, \alpha) \in (R^n \times \mathbb{R}) \setminus \{(0,0)\}$ such that: $p \cdot x + \alpha t \leq p \cdot x' + \alpha t'$, $\forall (x,t) \in B, \forall (x', t') \in A$. Since B is a vector space, if $(x,t) \in B$ and if $(x',t') \in A$, one has: $\forall \mu \in \mathbb{R}$, $p \cdot \mu x + \alpha \mu t \leq p \cdot x' + \alpha t'$. Hence $0 = p \cdot x + \alpha t \leq p \cdot x' + \alpha t'$, $\forall (x,t) \in B, \forall (x',t') \in A$. Let r verify : $0 = h(0) \leq f(0) < r$. One has $(0, r) \in A$. Hence $\alpha r \geq 0$, which implies $\alpha \geq 0$. If $\alpha = 0$, then $p \cdot x' \geq 0$, $\forall x' \in U$. Since U is open and contains 0, one has $p = 0$. That contradicts $(p, \alpha) \neq (0,0)$. Thus $\alpha > 0$.

Define a linear functional h_1 on $\mathbb{R}^n$ by $h_1(x) = -(p \cdot x / \alpha)$. Since $(x, h(x)) \in B$ if $x \in M$, then for all x in M, $p \cdot x + \alpha h(x) = 0$. Therefore, h_1 coincides with h on M. Let $x \in \mathrm{i}(\mathrm{dom}(f))$ and $\varepsilon > 0$. Since $(x, f(x) + \varepsilon) \in A$, one has $p \cdot x + \alpha(f(x) + \varepsilon) \geq 0$. That means that $h_1(x) \leq f(x) + \varepsilon$, $\forall \varepsilon > 0$, which proves that $h_1(x) \leq f(x)$, $\forall x \in \mathrm{i}(\mathrm{dom}(f))$. Now let $x \in \mathrm{dom}(f)$. For any $\lambda \in]0,1[$, one has $\lambda x \in U$ since $0 \in U$. From the previous result, one gets:

$$h_1(\lambda x) \leq f(\lambda x) \leq \lambda f(x) + (1 - \lambda) f(0).$$

Let λ tend to 1. We obtain that $h_1(x) \leq f(x)$. To sum up, we have proved that there exists a linear extension h_1 of h which verifies $\forall x \in \mathbb{R}^n$, $h_1(x) \leq f(x)$.

Let us now prove the general case. Let F denotes the vector subspace (since $0 \in \mathrm{ri}(\mathrm{dom}(f))$) affine hull of $\mathrm{dom}(f)$. Though the previous case has been formulated in $\mathbb{R}^n$, it can be applied in any finite-dimensional vector space. In particular, if we consider the restriction of the function f to the space F and the vector space $N = M \cap F$, we can deduce the existence of a linear extension $\tilde{h}$ of h on F which verifies $\forall x \in F$, $\tilde{h}(x) \leq f(x)$.

Let $F^\perp$ denote the orthogonal subspace to F. Recall that $\mathbb{R}^n$ is the direct sum of F and $F^\perp$. Let h_1 be the linear functional on $\mathbb{R}^n$ defined by

$$h_1(x) = \begin{cases} \tilde{h}(x) & \text{if } x \in F; \\ 0 & \text{if } x \in F^\perp. \end{cases}$$

Equivalently, $h_1(x) = (\tilde{h} \circ \pi)(x)$, where π is the orthogonal projection on F. Obviously, h_1 extends $\tilde{h}$ and consequently h. Since if $x \notin F$, then $f(x) = +\infty$, we deduce: $\forall x \in \mathbb{R}^n$, $\tilde{h}(x) \leq f(x)$. □

We can emphasize that the assumption that $0 \in \text{ri}(\text{dom}(f))$ is a costless assumption, since one can get it by translation. Formally, we state the following corollary.

Corollary 6.1.1. *Let f be a convex function from $\mathbb{R}^n$ into $\mathbb{R} \cup \{+\infty\}$. Let V be an affine subset of $\mathbb{R}^n$ that intersects* ri(dom(f)). *Let g be an affine functional on V such that:*

$$\forall x \in V, \ g(x) \leq f(x).$$

Then there exists an affine extension g_1 of g on $\mathbb{R}^n$ which verifies:

$$\forall x \in \mathbb{R}^n, \ g_1(x) \leq f(x).$$

Proof. Let $\overline{x} \in V \cap \text{ri}(\text{dom}(f))$. We define $M = V - \{\overline{x}\}$, the linear subspace of $\mathbb{R}^n$ parallel to V, $f_1 : \mathbb{R}^n \to \mathbb{R} \cup \{+\infty\}$ a convex function by $f_1(x) = f(x + \overline{x})$, h a linear functional by $h(x) = g(x + \overline{x})$. Using Theorem 6.1.1, we get the existence of a linear extension h_1 of h on $\mathbb{R}^n$ which verifies: $\forall x \in \mathbb{R}^n$, $h_1(x) \leq f_1(x)$. One can check easily that by translation, the affine functional defined on $\mathbb{R}^n$ by $g_1(x) = h_1(x - \overline{x})$ satisfies the required inequality. □

The Hahn-Banach theorem has several useful consequences (see for example Exercise 6.15). It implies in particular any of the separation theorems stated in Chapter 2. In what follows, we use the Hahn-Banach theorem to deduce the subdifferentiability theorem (Proposition 6.5.3).

6.2 Sublinear functions

Definition 6.2.1. *Let f be a function from $\mathbb{R}^n$ into $\mathbb{R} \cup \{+\infty\}$. We say that f is sublinear if*

$$\forall x \in \mathbb{R}^n, \forall y \in \mathbb{R}^n, \forall t > 0, \ f(x+y) \leq f(x) + f(y) \text{ and } f(tx) = tf(x).$$

It is worth noticing that f is sublinear if, and only if, f is convex, and *positively homogeneous* (i.e. $\forall t > 0, \forall x \in \mathbb{R}^n$, $f(tx) = tf(x)$).

Proposition 6.2.1. *Let $f : \mathbb{R}^n \to \mathbb{R} \cup \{+\infty\}$ be a sublinear function. If $0 \in \mathrm{dom}(f)$ then $f(0) = 0$ and $\forall x \in \mathbb{R}^n$, $f(x) + f(-x) \geq 0$.*

Proof. Left for the reader (see Exercise 6.1). □

For any sublinear function f, let us define:

$$\Delta(f) = \{m \in \mathbb{R}^n \mid m \cdot x \leq f(x),\ \forall x \in \mathbb{R}^n\}.$$

Proposition 6.2.2. *(i) Let f be a sublinear function from $\mathbb{R}^n$ into $\mathbb{R}$. Then $\Delta(f)$ is nonempty. Moreover, for any $y \in \mathbb{R}^n$, for any $z \in [-f(-y), f(y)]$, there exists $m \in \Delta(f)$ which verifies $m \cdot y = z$.*
(ii) Let f, g be two sublinear functions from $\mathbb{R}^n$ into $\mathbb{R}$ which verify $f \leq g$. Then $\Delta(f) \subset \Delta(g)$.
(iii) If $\lambda > 0$ and if f is sublinear, then $\Delta(\lambda f) = \lambda \Delta(f)$.

Proof. (i) We first prove the second claim with $y \neq 0$. Let $z \in [-f(-y), f(y)]$ and let D be the one-dimensional vector subspace of $\mathbb{R}^n$ generated by y. Define the linear functional h on D by $h(\lambda y) = \lambda z$ where $\lambda \in \mathbb{R}$. Let us check that $\forall x \in D, h(x) \leq f(x)$. Let $x = \lambda y$, with $\lambda \in \mathbb{R}$. If $\lambda > 0$, then $h(x) = \lambda z \leq \lambda f(y) = f(\lambda y) = f(x)$. If $\lambda = 0$, then $h(0) = 0 = f(0)$. If $\lambda < 0$, then $h(x) = \lambda z \leq -\lambda f(-y) = f(\lambda y) = f(x)$. From the Hahn-Banach theorem, there exists a linear functional h_1 extending h and verifying $h_1(x) \leq f(x), \forall x \in \mathbb{R}^n$. One can identify h_1 to a vector m of $\mathbb{R}^n$. It is obvious, from the very definition, that $m \in \Delta(f)$. It is also obvious $m \cdot y = h(y) = z$.

Now, if $y = 0$, one has $\forall m \in \mathbb{R}^n$, $m \cdot y = 0 = f(0) = z$. The proof is complete.
(ii) and (iii) The proofs are easy. □

Corollary 6.2.1. *Let f be a sublinear function from $\mathbb{R}^n$ into $\mathbb{R}$. Then $f(x) = \sup\{m \cdot x \mid m \in \Delta(f)\}$.*

Proof. Left for the reader (see Exercise 6.2). □

Proposition 6.2.3. *If f is sublinear from $\mathbb{R}^n$ into $\mathbb{R}$, then $\Delta(f)$ is convex, compact, nonempty.*

Proof. The set $\Delta(f)$ is nonempty by Proposition 6.2.2. It is obvious that $\Delta(f)$ is convex and closed. It suffices to show that it is bounded. Let $(e^1, \ldots, e^n)$ be a basis of $\mathbb{R}^n$. If $(m_1, \ldots, m_n)$ denote the coordinates of m in this basis, one has, for all $m \in \Delta(f) : \forall i, -f(-e^i) \leq m_i \leq f(e^i)$. □

From Proposition 6.2.3, we can consider Δ as a mapping from the set of real-valued sublinear functions to the set of nonempty convex compact subsets of $\mathbb{R}^n$.

6.3 Support functions

Definition 6.3.1. *Let C be a nonempty subset of $\mathbb{R}^n$. The support function $\delta_C^* : \mathbb{R}^n \to \mathbb{R} \cup \{+\infty\}$ of C is defined by:*

$$\forall x \in \mathbb{R}^n, \ \delta_C^*(x) = \sup\{x \cdot y \mid y \in C\}.$$

Proposition 6.3.1. *The support function of any nonempty set C of $\mathbb{R}^n$ is sublinear.*

Proof. See exercise 6.2. □

Proposition 6.3.2. *If C, C' are nonempty, then $\delta_{C+C'}^* = \delta_C^* + \delta_{C'}^*$. If C is nonempty and λ is strictly positive then $\delta_{\lambda C}^* = \lambda \delta_C^*$. Moreover, $\delta_C^* = \delta_{\mathrm{co}\, C}^*$*

Proof. See exercise 6.2. □

Proposition 6.3.3. *If C is a nonempty compact subset of $\mathbb{R}^n$, then δ_C^* is real-valued.*

Proof. The set C being nonempty and compact, the supremum of $x \cdot y$ on C is attained by some $y_C \in C$. Hence, $\delta_C^*(x) = x \cdot y_C < +\infty$. □

From Proposition 6.3.3, we can define a mapping δ^* from the set of nonempty, compact subsets of $\mathbb{R}^n$ into the set of real-valued sublinear functions on $\mathbb{R}^n$ by setting $\delta^*(C) = \delta_C^*$, where C is a compact, nonempty subset of $\mathbb{R}^n$. Let $S(\mathbb{R}^n, \mathbb{R})$ denote the set of sublinear functions from $\mathbb{R}^n$ into $\mathbb{R}$, and let $CK(\mathbb{R}^n)$ denote the set of nonempty convex and compact subsets of $\mathbb{R}^n$.

Proposition 6.3.4. *The mapping δ^* is a one-to-one mapping from $CK(\mathbb{R}^n)$ onto $S(\mathbb{R}^n, \mathbb{R})$. Its inverse mapping is Δ.*

Proof. One has $\Delta(\delta_C^*) = \{m \in \mathbb{R}^n \mid m \cdot x \leq \delta_C^*(x), \forall x \in \mathbb{R}^n\}$. It is obvious that $C \subset \Delta(\delta_C^*)$. Assume there exists $m \in \Delta(\delta_C^*)$ and $m \notin C$. From the second Separation theorem (Proposition 2.1.8), there exists $p \in \mathbb{R}^n$ and $\alpha \in \mathbb{R}$ such that $p \cdot y < \alpha < p \cdot m$, $\forall y \in C$. Hence $\delta_C^*(p) \leq \alpha < p \cdot m$, which leads to a contradiction since $m \in \Delta(\delta_C^*)$. Hence $C = \Delta(\delta_C^*)$, i.e., $C = \Delta(\delta^*(C))$. On the other hand, if f is sublinear, in view of Corollary 6.2.1, one has $\delta^*(\Delta(f)) = f$.

We have finally, $\forall C \in CK(\mathbb{R}^n)$, $\Delta(\delta^*(C)) = C$ and $\forall f \in S(\mathbb{R}^n, \mathbb{R})$, $\delta^*(\Delta(f)) = f$, which ends the proof. □

Remark 6.3.1. For any linear subspace F of $\mathbb{R}^n$, let us denote by $CK(F)$ the set of nonempty convex compact subsets of F and by $S(F, \mathbb{R})$, the set of sublinear, real-valued functions defined on F. We can define in a similar way a mapping $\Delta^F : S(F, \mathbb{R}) \to CK(F)$ as follows:

$$\Delta^F(g) = \{m \in F \mid m \cdot x \leq g(x), \ \forall x \in F\},$$

The proof that $\Delta^F(g)$ is nonempty convex and compact is an easy adaptation of the one given for Proposition 6.2.3.
Once again, in a similar way, let $\delta^{*F} : CK(F) \to S(F, \mathbb{R})$ be defined by

$$\forall x \in F, \forall C \in CK(F),\ \delta^{*F}(C)(x) = \sup\{x \cdot y \mid y \in C\}.$$

The reader will verify (see Exercise 6.4) that, as in the previous proposition, δ^{*F} is a one-to-one mapping from $CK(F)$ onto $S(F, \mathbb{R})$, with Δ^F as inverse mapping.

Proposition 6.3.5. *(i) Let $f, g \in S(\mathbb{R}^n, \mathbb{R})$. One has $\Delta(f+g) = \Delta(f) + \Delta(g)$.*
(ii) Let F be a linear subspace of $\mathbb{R}^n$, let $f, g \in S(F, \mathbb{R})$. One has $\Delta^F(f+g) = \Delta^F(f) + \Delta^F(g)$.

Proof. (i) One has by Proposition 6.3.4, $\delta^*(\Delta(f)) = f$, $\delta^*(\Delta(g)) = g$, $\delta^*(\Delta(f+g)) = f+g$, and by Proposition 6.3.2, $\delta^*(\Delta(f)+\Delta(g)) = \delta^*(\Delta(f)) + \delta^*(\Delta(g))$. Hence one obtains that $\Delta(\delta^*(\Delta(f) + \Delta(g))) = \Delta(\delta^*(\Delta(f+g)))$. From Proposition 6.3.4, one gets: $\Delta(f) + \Delta(g) = \Delta(f+g)$.
(ii) the proof is an easy adaptation of (i). □

6.4 Directional derivatives

Let $f : \mathbb{R}^n \to \mathbb{R} \cup \{+\infty\}$ be a function. Let x satisfy $f(x) < +\infty$. The *derivative* of f at x *in the direction of* $y \in \mathbb{R}^n$ is:

$$f'(x; y) = \lim_{\lambda > 0, \lambda \to 0} \frac{f(x + \lambda y) - f(x)}{\lambda}$$

when this limit exists. This limit may be equal to $+\infty$ or $-\infty$.

Proposition 6.4.1. *Let f be a convex function from $\mathbb{R}^n$ into $\mathbb{R} \cup \{+\infty\}$ and let $x \in \operatorname{dom}(f)$. Then we have:*
(i) $\lambda \geq \lambda' > 0 \Rightarrow (1/\lambda)(f(x+\lambda y) - f(x)) \geq (1/\lambda')(f(x + \lambda' y) - f(x))$.
Hence, $f'(x; y) = \inf_{\lambda > 0}(1/\lambda)(f(x + \lambda y) - f(x))$.
(ii) $f'(x; \mu y) = \mu f'(x; y)$, $\forall \mu > 0$; $f'(x; 0) = 0$; if $f'(x; y) > -\infty, f'(x; y') > -\infty$ then for any $\alpha \in]0, 1[$, we have

$$f'(x; \alpha y + (1-\alpha) y') \leq \alpha f'(x; y) + (1-\alpha) f'(x; y').$$

(iii) $-f'(x; -y) \leq f'(x; y)$, $\forall y \in \mathbb{R}^n$.

Proof. Define the convex function $h : \mathbb{R}^n \to \mathbb{R} \cup \{+\infty\}$ by $h(\lambda) = f(x + \lambda y) - f(x)$. Let $0 < \lambda' \leq \lambda$. We have: $h(\lambda') = h((\lambda'/\lambda)\lambda) \leq (\lambda'/\lambda) h(\lambda) + (1 - (\lambda'/\lambda)) h(0) = (\lambda'/\lambda) h(\lambda)$.

The statements (ii) and (iii) are let as exercises. □

In Assertion (ii) of Proposition 6.4.1, the condition that $f'(x;y) > -\infty$ and $f'(x;y') > -\infty$ is very important to obtain that for any $\alpha \in]0,1[$, we have $f'(x;\alpha y + (1-\alpha)y') \leq \alpha f'(x;y) + (1-\alpha)f'(x;y')$. Indeed, consider the function on $\mathbb{R}$ defined by $f(x) = 0, \forall x < 0, f(0) = 1, f(x) = +\infty, \forall x > 0$. This function is convex. We find that $f'(0;1) = +\infty, f'(0;-1) = -\infty, f'(0;0) = 0$. While $0 = (1/2)1 + (1/2)(-1)$, we cannot write $f'(0;0) \leq (1/2)f'(0;1) + (1/2)f'(0;-1)$ because the left-side member is meaningless.
We want to define now the partial derivative for a convex function $f : \mathbb{R}^n \rightarrow \mathbb{R} \cup \{+\infty\}$. The ith partial derivative of f at $x \in \mathbb{R}^n$, which is denoted by $(\partial f/\partial x_i)(x)$, is defined by:

$$\frac{\partial f}{\partial x_i}(x) = \lim_{\lambda \to 0} \frac{f(x + \lambda e^i) - f(x)}{\lambda} = \lim_{\lambda \to 0} \frac{f(x - \lambda e^i) - f(x)}{(-\lambda)}$$

where $\{e^1, \ldots, e^n\}$ is the the canonical basis of $\mathbb{R}^n$. Hence $\forall i$, $(\partial f/\partial x_i)(x) = f'(x; e^i) = -f'(x; -e^i)$.
We will denote by $f'(x;.)$ the function from $\mathbb{R}^n$ to $\overline{\mathbb{R}}$ defined by $y \rightarrow f'(x;y)$.

Proposition 6.4.2. *Assume that* $f : \mathbb{R}^n \rightarrow \mathbb{R} \cup \{+\infty\}$ *is convex. Let* $F = \text{aff}(\text{dom}(f)) - \{x_0\}$, *where* $x_0 \in \text{aff}(\text{dom}(f))$.
(i) The restriction to F *of the function* $f'(x;.)$ *is real-valued if, and only if,* $x \in \text{ri}(\text{dom}(f))$.
(ii) If $x \in \text{ri}(\text{dom}(f))$ *then* $f'(x;.)$ *is sublinear on* F.
(iii) The function $f'(x;.)$ *is real-valued on* $\mathbb{R}^n$ *if, and only if,* $x \in \text{i}(\text{dom}(f))$.

Proof. (i) First assume that $x \in \text{ri}(\text{dom}(f))$. We then prove that $f'(x;y)$ is real-valued for any $y \in F$. It suffices to check that $f'(x;y) < +\infty$, $\forall y \in F$. For $\lambda > 0$, sufficiently small, we have $x + \lambda y \in \text{dom}(f)$. Hence $f'(x;y) \leq (1/\lambda)(f(x + \lambda y) - f(x)) < +\infty$.
Conversely, assume $f'(x;y)$ is real-valued for every $y \in F$. We claim that $x \in \text{ri}(\text{dom}(f))$. If not, from Corollary 1.2.6, there exist $y \in F$ and a sequence of strictly positive numbers $(\lambda_n)_{n \in \mathbb{N}}$ converging to 0 such that $x + \lambda_n y \notin \text{dom}(f), \forall n$. In this case, we have $(f(x + \lambda_n y) - f(x))/\lambda_n = +\infty$, $\forall n$. From its very definition, $f'(x;y) = +\infty$ which contradicts $f'(x;y) < +\infty$.
(ii) It follows from Assertion (ii) of Proposition 6.4.1.
(iii) Apply the same argument as in (ii). □

6.5 Subgradients and subdifferential of a convex function

Let f be a convex function on $\mathbb{R}^n$. The element $x^* \in \mathbb{R}^n$ is called *subgradient* of f at x if $f(z) \geq f(x) + x^* \cdot (z - x)$, $\forall z \in \mathbb{R}^n$. The set of subgradients of f at x is called *subdifferential* of f at x and is noted $\partial f(x)$. If $\partial f(x)$ is nonempty, we say that f is *subdifferentiable at* x.

Proposition 6.5.1. *Let f be a convex function on $\mathbb{R}^n$. Let $x \in \operatorname{dom}(f)$. Then:*

$$x^* \in \partial f(x) \Leftrightarrow f'(x;y) \geq x^* \cdot y, \forall y \in \mathbb{R}^n.$$

In other words, $\Delta(f'(x;.)) = \partial f(x)$ if $x \in \operatorname{ri}(\operatorname{dom}(f))$.

Proof. Let $x^* \in \partial f(x)$. $\forall y \in \mathbb{R}^n, \forall \lambda > 0$, one has:

$$f(x+\lambda y) - f(x) \geq x^* \cdot (\lambda y) = \lambda x^* \cdot y.$$

Hence $f'(x;y) \geq x^* \cdot y$.
Conversely, assume that $f'(x;y) \geq x^* \cdot y, \forall y \in \mathbb{R}^n$. Let $z \in \mathbb{R}^n$. Define $y = z - x$. One has by Assertion (i) of Proposition 6.4.1, $f(x+y) - f(x) \geq f'(x;y)$, i.e. $f(z) - f(x) \geq x^* \cdot (z-x)$. That means that $x^* \in \partial f(x)$. □

Proposition 6.5.2. *Let $f : \mathbb{R}^n \to \mathbb{R} \cup \{+\infty\}$ be a convex function. Then*
(i) $\partial f(x)$ is convex and closed.
(ii) If $x \notin \operatorname{dom}(f)$, then $\partial f(x) = \emptyset$.
(iii) The subdifferential of f, $\partial f(x)$, is compact, nonempty if, and only if, $x \in \operatorname{i}(\operatorname{dom}(f))$.

Proof. (i) It is easy to check that $\partial f(x)$ is convex and closed.
(ii) Let $z \in \operatorname{dom}(f)$. If $x \notin \operatorname{dom}(f)$ then $f(x) = +\infty$. There exists no $x^* \in \mathbb{R}^n$ such that $x^* \cdot (z-x) + f(x) \leq f(z) < +\infty$.
(iii) We first prove that if $x \in \operatorname{i}(\operatorname{dom}(f))$ then $\partial f(x)$ is compact and nonempty. Indeed, if $x \in \operatorname{i}(\operatorname{dom}(f))$ then from Proposition 6.4.2, the function $\mathbb{R}^n \to \overline{\mathbb{R}}$ defined by $y \to f'(x;y)$ is real-valued and sublinear. From Proposition 6.2.3, $\Delta(f'(x;.))$ is compact, nonempty. Hence $\partial f(x)$ is compact and nonempty by Proposition 6.5.1.

Conversely, let us assume $\partial f(x)$ nonempty and compact; we want to prove that $x \in \operatorname{i}(\operatorname{dom}(f))$. Assume the contrary, i.e. x is in the boundary of $\operatorname{dom}(f)$. From Corollary 2.1.1, there exists $p \in \mathbb{R}^n \setminus \{0\}$ such that $p \cdot x \geq p \cdot z$, $\forall z \in \operatorname{dom}(f)$. Let $x^* \in \partial f(x)$. We claim that $x^* + \lambda p \in \partial f(x)$, $\forall \lambda \geq 0$. Indeed, we have, for $z \in \operatorname{dom}(f)$:

$$f(z) - f(x) \geq x^* \cdot (z-x) \geq (x^* + \lambda p) \cdot (z-x), \ \forall \lambda \geq 0.$$

If $z \notin \operatorname{dom}(f))$, then $f(z) = +\infty$ and the previous inequality still holds. Thus our claim is true. But that implies that $\partial f(x)$ is not bounded in contradiction with the assumption that it is compact. Hence $x \in \operatorname{i}(\operatorname{dom}(f))$. □

Proposition 6.5.3 (Subdifferentiability theorem). *Let $f : \mathbb{R}^n \to \mathbb{R} \cup \{+\infty\}$ be a convex function such that $\operatorname{dom}(f) \neq \emptyset$. If $x \in \operatorname{ri}(\operatorname{dom}(f))$, we have:*
(i) $\partial f(x) \neq \emptyset$ and
(ii) for every $y \in \mathbb{R}^n$, we have $f'(x;y) = \sup\{p \cdot y \mid p \in \partial f(x)\}$.

Proof. (i) Let $x \in \mathrm{ri}(\mathrm{dom}(f))$. Consider $F = \mathrm{aff}(\mathrm{dom}(f)) - \{x\}$. The set F is a vector subspace. It is easy to check that if $F = \{0\}$ (i.e. $\mathrm{dom}(f) = \{x\}$) then $\partial f(x) = \mathbb{R}^n$ and the statement is proved. Assume $F \neq \{0\}$ and choose some $y \in F$ which satisfies $y \neq 0$. Let g be the convex function $\mathbb{R}^n \to \mathbb{R} \cup +\infty$ defined by $g(z) = f(x+z) - f(x)$. We have $g(\lambda y) = f(x + \lambda y) - f(x)$. Define on the linear subspace generated by y, the linear functional h by $h(\lambda y) = \lambda f'(x;y)$. One has: if $\lambda > 0$ then $g(\lambda y) \geq h(\lambda y)$. If $\lambda = 0$, then $g(0) = h(0) = 0$. If $\lambda < 0$, then in view of Assertion (iii) of Proposition 6.4.1, $g(\lambda y) \geq -\lambda f'(x;-y) \geq \lambda f'(x;y) = h(\lambda y)$. From the Hahn-Banach theorem (Theorem 6.1.1), h can be extended to a linear functional h_1 on $\mathbb{R}^n$ verifying $h_1(z) \leq f(x+z) - f(x)$, $\forall z \in \mathbb{R}^{\cdot}$. Let $x^* \in \mathbb{R}^n$ such that $h_1(z) = x^* \cdot z$. Clearly, the vector x^* belongs to $\partial f(x)$.
(ii) If $\mathrm{i}(\mathrm{dom}(f)) \neq \emptyset$, then if $x \in \mathrm{i}(\mathrm{dom}(f))$, we first have that $f'(x;.)$ is sublinear, real-valued by Assertion (iii) of Proposition 6.4.2. From Proposition 6.5.1, $\Delta(f'(x;.)) = \partial f(x)$. Thus, by Proposition 6.3.4,

$$\forall y \in \mathbb{R}^n,\ f'(x;y) = \delta^*(\partial f(x))(y) = \sup\{x^* \cdot y \mid x^* \in \partial f(x)\}.$$

Consider now the case where $\mathrm{i}(\mathrm{dom}(f)) = \emptyset$. Let $F^{\perp}$ denote the orthogonal subspace to F; recall that $\mathbb{R}^n$ is the direct sum of F and $F^{\perp}$. Let us first remark that for any

$$\forall x^* \in \partial f(x), \forall v \in F^{\perp}, \forall \lambda \geq 0,\ x^* + \lambda v \in \partial f(x).$$

Indeed, if $z \notin \mathrm{dom}(f)$, then $f(z) = +\infty$ and $f(z) \geq f(x) + (h + \lambda v) \cdot (z - x)$. If $z \in \mathrm{dom}(f)$ then $z - x \in F$ and consequently $v \cdot (z - x) = 0$. In both cases, we have

$$f(z) \geq f(x) + (x^* + \lambda v) \cdot (z - x)$$

If $y \notin F$, let us prove that $f'(x;y) = \sup\{x^* \cdot y \mid x^* \in \partial f(x)\} = +\infty$. We write $y = y_F + y_{F^{\perp}}$ with $y_{F^{\perp}} \neq 0$. Using Assertion (i) and the previous remark, for any $x^* \in \partial f(x)$ and every $\lambda \geq 0$, $x^* + \lambda y_{F^{\perp}} \in \partial f(x)$. But $(x^* + \lambda y_{F^{\perp}}) \cdot y = x^* \cdot y + \lambda \|y_{F^{\perp}}\|^2$. The last term converges to $+\infty$ when λ tends to $+\infty$. That implies $\sup\{x^* \cdot y \mid x^* \in \partial f(x)\} = +\infty$. We also have $\forall \lambda > 0$, $f(x + \lambda y) = +\infty$ since $x + \lambda y \notin \mathrm{dom}(f)$. Hence $f'(x;y) = +\infty$ by Assertion (i) of Proposition 6.4.1.

If $y \in F$, we will then prove that $f'(x;y) = \sup\{x^* \cdot y \mid x^* \in \partial f(x)\} < +\infty$. The fact that $f'(x;y) < +\infty$ follows from Assertion (i) of Proposition 6.4.2. In order to to prove the remaining equality, let j denote the restriction to F of $f'(x;.)$. In view of Assertion (ii) of Proposition 6.4.2, since $x \in \mathrm{ri}(\mathrm{dom}(f))$, we have $j \in S(F, \mathbb{R})$. We will now use the notations of Remark 6.3.1.

Claim. $\Delta^F(j) = \partial f(x) \cap F$; hence $\partial f(x) \cap F$ is compact.

Recall that $\Delta^F(j) = \{m \in F \mid m \cdot z \leq j(z),\ \forall z \in F\}$. Since $f'(x;z) \leq f(x+z) - f(x)$, we have $m \cdot z \leq f(x+z) - f(x)$, $\forall m \in \Delta^F(j), \forall z \in F$. If $z \notin F$, the last inequality still holds since $f(x+z) = +\infty$ and $f(x) < +\infty$. We have thus proved that $\Delta^F(j) \subset \partial f(x) \cap F$. To prove the reverse inclusion, take some $x^* \in \partial f(x) \cap F$ and some $z \in F$. For $\lambda > 0$, we have $f(x+\lambda z) - f(x) \geq x^* \cdot (\lambda z)$, and thus $f'(x;y) \geq x^* \cdot y$. Since z is arbitrarily chosen in F, we actually have $x^* \in \Delta^F(j)$. □

Thus

$$f'(x;y) = j(y) = (\delta^{*F}(\Delta^F(j)))(y) = (\delta^{*F}(\partial f(x) \cap F))(y),$$

or

$$f'(x;y) = \sup\{x^* \cdot y \mid x^* \in \partial f(x) \cap F\} \leq \sup\{x^* \cdot y \mid x^* \in \partial f(x)\}.$$

Since, in view of Proposition 6.5.1, $f'(x;y) \geq x^* \cdot y$, $\forall x^* \in \partial f(x)$, then $f'(x;y) = \sup\{p \cdot y \mid y \in \partial f(x)$. The proof is now complete. □

It is worth noticing, from the proof of the previous proposition, that we have for $x \in \mathrm{ri}(\mathrm{dom}(f))$:
(i) $\partial f(x) \cap F$ is compact, where $F = \mathrm{aff}(\mathrm{dom}(f)) - \{x\}$,
(ii) if $x^* \in \partial f(x)$ then $\forall v \in F^\perp$, $x^* + v \in \partial f(x)$. Hence $\partial f(x) = (\partial f(x) \cap F) + F^\perp$.
But we have actually:

Proposition 6.5.4. *Let $f : \mathbb{R}^n \to \mathbb{R} \cup \{+\infty\}$ be a convex. Then $x \in \mathrm{ri}(\mathrm{dom}(f))$ if, and only if, $\partial f(x) \cap F$ is compact, nonempty, where $F = \mathrm{aff}(\mathrm{dom}(f)) - \{x\}$.*

Proof. In the proof of the previous proposition, we have shown that if $x \in \mathrm{ri}(\mathrm{dom}(f)$, then $\partial f(x) \cap F$ is compact, nonempty.

To prove the converse, assume that $x \notin \mathrm{ri}(\mathrm{dom}(f))$ while $\partial f(x) \cap F$ is compact, nonempty. In this case x belongs to the relative boundary of $\mathrm{dom}(f)$. Using Corollary 2.1.4, there exists $p \in F$, $p \neq 0$ such that $p \cdot x \geq p \cdot y$, $\forall y \in \mathrm{dom}(f)$. One can easily check that $\forall \lambda \geq 0, \forall x^* \in \partial f(x) \cap F$, $x^* + \lambda p \in \partial f(x) \cap F$ and, in this case, $\partial f(x) \cap F$ is not bounded. That is a contradiction with the assumption that $\partial f(x) \cap F$ is compact. □

Lemma 6.5.1. *Let $f : \mathbb{R}^n \to \mathbb{R} \cup \{+\infty\}$ be a convex function. Then*

$$\mathrm{ri}(\mathrm{epi}(f)) = \{(x, \mu) \in \mathbb{R}^n \times \mathbb{R} \mid x \in \mathrm{ri}(\mathrm{dom}(f)), \mu > f(x)\}.$$

Proof. Let us prove that $\mathrm{ri}(\mathrm{epi}(f)) \subset \{(x, \mu) \in \mathbb{R}^n \times \mathbb{R} \mid x \in \mathrm{ri}(\mathrm{dom}(f)), \mu > f(x)\}$. Let $z_0 = (x_0, \mu_0) \in \mathrm{ri}(\mathrm{epi}(f))$ and let $z = (x, \mu) \in \mathrm{epi}(f)$ with $\mu > f(x)$. There exists $\lambda > 1$ such that $(1-\lambda)z + \lambda z_0 \in \mathrm{epi}(f)$, i.e. $x' = (1-\lambda)x + \lambda x_0 \in \mathrm{dom}(f)$ and $f(x') \leq (1-\lambda)\mu + \lambda\mu_0$. One can write: $x_0 = (1-(1/\lambda))x + (1/\lambda)x'$. Hence:

$$\begin{aligned} f(x_0) &\leq (1-(1/\lambda))f(x) + (1/\lambda)f(x') \\ &< (1-(1/\lambda))\mu + (1/\lambda)f(x') \\ &\leq (1-1/\lambda)\mu + (1/\lambda)((1-\lambda)\mu + \lambda\mu_0) = \mu_0 \end{aligned}$$

Since x may be arbitrarily chosen in dom(f), one has $x_0 \in \mathrm{ri}(\mathrm{dom}(f))$ (Corollary 1.2.5).

We now prove the reverse inclusion, i.e.

$$\{(x,\mu) \in \mathbb{R}^n \times \mathbb{R} \mid x \in \mathrm{ri}(\mathrm{dom}(f)), \mu > f(x)\} \subset \mathrm{ri}(\mathrm{epi}(f)).$$

Let (x_0, μ_0) be such that $x_0 \in \mathrm{ri}(\mathrm{dom}(f))$ and $\mu_0 > f(x_0)$. We will prove that $(x_0, \mu_0) \in \mathrm{ri}(\mathrm{epi}(f))$. Assume the contrary. In view of Corollary 1.2.5, there exists $(x,\mu) \in \mathrm{epi}(f)$ such that $\forall \lambda > 1$, $(x,\mu) + \lambda((x_0,\mu_0) - (x,\mu)) \notin \mathrm{epi}(f)$, i.e.

$$\forall \lambda > 1, \; f(x + \lambda(x_0 - x) > \mu + \lambda(\mu_0 - \mu). \tag{6.1}$$

Since $x_0 \in \mathrm{ri}(\mathrm{dom}(f))$, there exists $\overline{\lambda}$ such that $\forall \lambda \in \left]1, \overline{\lambda}\right[$, $x + \lambda(x_0 - x) \in \mathrm{dom}(f)$. Recall that the restriction of f to aff(dom(f)) is continuous at x_0 (Corollary 5.2.3). Letting λ tend to 1 in (6.1), one gets $f(x_0) \geq \mu_0$, which contradicts $\mu_0 > f(x_0)$. □

Proposition 6.5.5. *Let $f_1, f_2, \ldots, f_p$ be p convex functions from $\mathbb{R}^n$ into $\mathbb{R} \cup \{+\infty\}$. If $\cap_{i=1}^p \mathrm{ri}(\mathrm{dom}(f_i)) \neq \emptyset$ then $\forall x \in \mathbb{R}^n, \partial(f_1 + f_2 + \ldots + f_p)(x) = \partial f_1(x) + \ldots + \partial f_p(x)$.*

Proof. We will first prove the proposition with $p = 2$.

It is obvious that $\partial f_1(x) + \partial f_2(x) \subset \partial(f_1 + f_2)(x)$. We now prove the converse. Let $\overline{x} \in \mathbb{R}^n$ and $\overline{x}^* \in \partial(f_1 + f_2)(\overline{x})$. We have here $\mathrm{dom}(f_1) \cap \mathrm{dom}(f_2) = \mathrm{dom}(f_1 + f_2)$ and thus, from Proposition 6.5.2, $\overline{x} \in \mathrm{dom}(f_1) \cap \mathrm{dom}(f_2)$. Consider the convex functions g_1 and g_2 :

$$\begin{aligned} g_1(x) &= f_1(x + \overline{x}) - f_1(\overline{x}) - x \cdot \overline{x}^*, \\ g_2(x) &= f_2(x + \overline{x}) - f_2(\overline{x}). \end{aligned}$$

One has $g_1(0) = g_2(0) = 0$, $\mathrm{dom}(g_1) = \mathrm{dom}(f_1) - \{\overline{x}\}$, $\mathrm{dom}(g_2) = \mathrm{dom}(f_2) - \{\overline{x}\}$ and $g_1(x) + g_2(x) \geq 0$, $\forall x \in \mathbb{R}^n$. Let $C_1 = \mathrm{epi}(g_1)$ and $C_2 = \{(x,\mu) \in \mathbb{R}^n \times \mathbb{R} \mid (x, -\mu) \in \mathrm{epi}(g_2)\}$. From the previous lemma, $\mathrm{ri}(C_1) = \{(x,\mu) \in \mathbb{R}^n \times \mathbb{R} \mid x \in \mathrm{ri}(\mathrm{dom}(g_1)), \mu > g_1(x)\}$, and $\mathrm{ri}(C_2) = \{(x,\mu) \in \mathbb{R}^n \times \mathbb{R} \mid x \in \mathrm{ri}(\mathrm{dom}(g_2)), \mu < -g_2(x)\}$. One has $\mathrm{ri}(C_1) \cap \mathrm{ri}(C_2) = \emptyset$ or equivalently $0 \notin \mathrm{ri}(C_1) - \mathrm{ri}(C_2)$. Since $\mathrm{ri}(C_1 - C_2) = \mathrm{ri}(C_1) - \mathrm{ri}(C_2)$ (see Exercise 1.27), one deduces from Corollary 2.1.4 that there exists $(p,\alpha) \in \mathbb{R}^n \times \mathbb{R}, (p,\alpha) \neq (0,0)$ such that

$$0 < p \cdot x_1 + \alpha t_1 - p \cdot x_2 - \alpha t_2, \; \forall (x_1,t_1) \in \mathrm{ri}(C_1), \forall (x_2,t_2) \in \mathrm{ri}(C_2). \tag{6.2}$$

If $\alpha = 0$, (6.2) gives $0 < p \cdot x_1 - p \cdot x_2$, $\forall x_1 \in \mathrm{ri}(\mathrm{dom}(g_1)), \forall x_2 \in \mathrm{ri}(\mathrm{dom}(g_2))$. Since $\mathrm{ri}(\mathrm{dom}(g_1)) \cap \mathrm{ri}(\mathrm{dom}(g_2)) \neq \emptyset$, one gets a contradiction. Hence $\alpha \neq 0$. By continuity (6.2) gives

$$0 \leq p \cdot x_1 + \alpha t_1 - p \cdot x_2 - \alpha t_2, \; \forall (x_1,t_1) \in C_1, \forall (x_2,t_2) \in C_2. \tag{6.3}$$

Taking $x_1 = 0, t_1 > 0, x_2 = 0, t_2 < 0$, one obtains from (6.3) that $\alpha > 0$, since we have previously proved that $\alpha \neq 0$. We also have by (6.3) $p \cdot x_1 + \alpha t_1 \geq 0, \forall (x_1, t_1) \in C_1$. Hence

$$-(p/\alpha) \cdot x_1 \leq g_1(x_1) = f_1(x_1 + \overline{x}) - f_1(\overline{x}) - x_1 \cdot \overline{x}^*, \forall x_1 \in \mathbb{R}^n,$$

because, either $x_1 \in \text{dom}(g_1)$ and this inequality holds by taking $t_1 = g_1(x_1)$ in the inequality given just before, or $x_1 \notin \text{dom}(g_1)$ and the inequality is true since $g_1(x_1) = +\infty$. That means that $-(p/\alpha) + \overline{x}^* \in \partial f_1(\overline{x})$. By the same way one finds $(p/\alpha) \in \partial f_2(\overline{x})$. Finally, $\overline{x}^* = (-(p/\alpha) + \overline{x}^*) + (p/\alpha) \in \partial f_1(\overline{x}) + \partial f_2(\overline{x})$. The result has been proved for $p = 2$.

We can now proceed by induction. Since $\text{dom}(f_1 + f_2) = \text{dom}(f_1) \cap \text{dom}(f_2)$, we have

$$\text{ri}(\text{dom}(f_1 + f_2)) = \text{ri}(\text{dom}(f_1) \cap \text{dom}(f_2)) = \text{ri}(\text{dom}(f_1)) \cap \text{ri}(\text{dom}(f_2))$$

(the second equality holds because $\text{ri}(\text{dom}(f_1)) \cap \text{ri}(\text{dom}(f_2)) \neq \emptyset$, see Exercise 1.26). Since $\text{ri}(\text{dom}(f_1)) \cap \text{ri}(\text{dom}(f_2)) \cap \text{ri}(\text{dom}(f_3)) \neq \emptyset$, from the previous result we get:
$\partial(f_1 + f_2 + f_3)(x) = \partial(f_1 + f_2)(x) + \partial f_3(x) = \partial f_1(x) + \partial f_2(x) + \partial f_3(x)$, and so on. □

Remark 6.5.1. If $\lambda > 0$ then we have $\partial(\lambda f)(x) = \lambda \partial f(x)$, $\forall x \in \mathbb{R}^n$. This formula is also valid when $\lambda = 0$ provided that $\partial f(x) \neq \emptyset$. When $\partial f(x)$ is empty, we define $\lambda \partial f(x) = \{0\}$. With this rule and the one given in Chapter 5 for (λf), we always have $\partial(\lambda f)(x) = \lambda \partial f(x)$, $\forall x \in \mathbb{R}^n, \forall \lambda \geq 0$.

6.6 Differentiability of convex functions

Let $f : \mathbb{R}^n \to \mathbb{R} \cup \{+\infty\}$ be a function and $x \in \mathbb{R}^n$such that $f(x) \in \mathbb{R}$. The function f is *differentiable* at x if there exists $x^* \in \mathbb{R}^n$ (such a x^* will be unique) such that $\lim_{z \to x}(f(z) - f(x) - x^* \cdot (z - x))/\|z - x\| = 0$. Such a x^* is called *gradient* of f at x and noted $\nabla f(x)$.
Assume that f is differentiable at x. Then, for all $y \neq 0$, one has

$$\lim_{\lambda > 0, \lambda \to 0} \frac{f(x + \lambda y) - f(x) - \nabla f(x) \cdot \lambda y}{\lambda \|y\|} = 0$$

or $\big(f'(x; y) - \nabla f(x) \cdot y\big)/\|y\| = 0$. Hence $f'(x; y) = \nabla f(x) \cdot y$, $\forall y \in \mathbb{R}^n$. In particular, for $i = 1, \ldots, n$, $\nabla f(x) \cdot e^i = \lim_{\lambda \to 0}\big(f(x + \lambda e^i) - f(x)\big)/\lambda = (\partial f / \partial x_i)(x)$, where e^i is the unit vector of the basis of $\mathbb{R}^n$.
Hence $\nabla f(x) = \big((\partial f / \partial x_1)(x), \ldots, (\partial f / \partial x_n)(x)\big)$ and if $y = (y_1, \ldots, y_n)$, then $f'(x; y) = \sum_{i=1}^n y_i (\partial f / \partial x_i)(x)$.

Proposition 6.6.1. *Let f be a convex function from $\mathbb{R}^n$ into $\mathbb{R} \cup \{+\infty\}$ and let $x \in \text{dom}(f)$. Then, f is differentiable at x if, and only if, there exists $x^* \in \mathbb{R}^n$ such that $\forall y \in \mathbb{R}^n, f'(x; y) = x^* \cdot y$. In this case, we have $x \in \text{i}(\text{dom}(f)), \nabla f(x) = x^*$.*

Proof. If f is differentiable at x, we have proved above that $\forall y \in \mathbb{R}^n$, $f'(x;y) = \nabla f(x) \cdot y$. From Proposition 6.4.2, Assertion (iii), $x \in \mathrm{i}(\mathrm{dom}(f))$.

Conversely, assume that $\exists x^* \in \mathbb{R}^n, \forall y \in \mathbb{R}^n, f'(x;y) = x^* \cdot y$. We have $x \in \mathrm{i}(\mathrm{dom}(f))$ by Assertion (iii) of Proposition 6.4.2. Define the convex function $g : \mathbb{R}^n \to \mathbb{R} \cup \{+\infty\}$ by $g(y) = f(x+y) - f(x) - x^* \cdot y$. One has $g(y) \geq 0$, $\forall y \in \mathbb{R}^n, g(0) = 0$ and $\lim_{\lambda>0,\lambda\to 0}(g(\lambda y)/\lambda) = 0$. If $y \neq 0$, then $(y/\|y\|)$ belongs to the unit-sphere which is included in the hypercube $[-1,1]^n$. From the Krein-Milman theorem (Proposition 3.1.4), there exist numbers $\beta_i(y)$ verifying $\sum_{i\in I} \beta_i(y) = 1, \beta_i(y) \geq 0$, $\forall i \in I$, such that $(y/\|y\|) = \sum_{i\in I} \beta_i(y) a_i$, where I is the set of indices of extreme points $\{a_i \mid i \in I\}$ of $[-1,1]^n$. One has $g(y) = g(\|y\|(y/\|y\|)) \leq \sum_{i\in I} \beta_i(y) g(\|y\| a_i)$. Hence $0 \leq (g(y)/\|y\|) \leq \sum_{i\in I}(g(\|y\|a_i)/\|y\|)$. But $\sum_{i\in I}(g(\|y\|a_i)/\|y\|)$ converges to 0 when $\|y\|$ tends to 0. In other words, $x^* = \nabla f(x)$ and the proof is complete. □

Proposition 6.6.2. *Let f be a convex function on $\mathbb{R}^n$. Then f is differentiable at x if, and only if $\partial f(x) = \{x^*\}$. In this case $\nabla f(x) = x^*$.*

Proof. Assume f differentiable at x. From the previous result, $\forall y \in \mathbb{R}^n$, $f'(x;y) = \nabla f(x) \cdot y$. From Proposition 6.5.1, we have: $x^* \in \partial f(x) \Leftrightarrow \nabla f(x) \cdot y \geq x^* \cdot y$, $\forall y \in \mathbb{R}^n$. That is possible if, and only if, $x^* = \nabla f(x)$.

Conversely, assume that $\partial f(x) = \{x^*\}$. From Assertion (iii) of Proposition 6.5.2, $x \in \mathrm{i}(\mathrm{dom}(f))$ and by Assertion (ii) of Proposition 6.5.3, $\forall y \in \mathbb{R}^n$, $f'(x;y) = x^* \cdot y$. From Proposition 6.6.1, f is differentiable at x and $\nabla f(x) = x^*$. □

Proposition 6.6.3. *Let f, g be two convex functions from $\mathbb{R}^n$ into $\mathbb{R}$. Assume that $\forall x \in \mathbb{R}^n$, $f(x) \geq g(x)$ and $f(x_0) = g(x_0)$. We assume that f is differentiable at x_0. Then g is also differentiable at x_0 and $\nabla f(x_0) = \nabla g(x_0)$.*

Proof. Since $\mathrm{dom}(g)$ is $\mathbb{R}^n$, $\partial g(x) \neq \emptyset$, $\forall x$ by Proposition 6.5.3. Let $x^* \in \partial g(x_0)$. One has: $f(x) - f(x_0) \geq g(x) - g(x_0) \geq x^* \cdot (x - x_0)$, $\forall x \in \mathbb{R}^n$. Hence $x^* \in \partial f(x_0)$ and $\partial g(x_0) \subset \partial f(x_0)$. Since f is differentiable at x_0, $\partial f(x_0) = \{\nabla f(x_0)\}$ (Proposition 6.6.2). Hence $\partial g(x_0) = \{\nabla f(x_0)\}$ and consequently, g is differentiable at x_0. □

An application of the previous proposition is the following version of the *Envelop Theorem.*

Proposition 6.6.4 (Envelop theorem). *Let f be a real-valued convex function on $\mathbb{R}^p \times \mathbb{R}^q$. For every $t \in \mathbb{R}^q$, consider the problem $\min\{f(x,t) \mid x \in K\}$, where K is a nonempty convex, compact set of $\mathbb{R}^p$. Define the function $V : \mathbb{R}^q \to \mathbb{R}$ by $V(t) = \min\{f(x,t) \mid x \in K\}$. Then V is real-valued and convex on $\mathbb{R}^q$.*

Suppose that $\forall x \in K$, the function $\mathbb{R}^q \to \mathbb{R}$ defined by $t \to f(x,t)$ is differentiable at $t_0 \in \mathbb{R}^q$. Then V is differentiable at t_0 and $\nabla V(t_0) = f'_2(x(t_0), t_0)$, where $x(t_0)$ is a minimizer, i.e. $f(x(t_0), t_0) = \min\{f(x,t_0) \mid x \in K\}$ and $f'_2(x(t_0), t_0)$ is the value at $(x(t_0), t_0)$ of the derivative of f with respect to t.

Proof. First, V is real-valued. Indeed, since K is compact and since for every $t \in \mathbb{R}^q$, the function $f(.,t)$ is continuous (as convex, real-valued, Theorem 5.2.1), there exists $x(t) \in K$ such that $V(t) = \min\{f(x,t) \mid x \in K\} = f(x(t),t) \in \mathbb{R}$.

We now prove that V is convex. Let $t \in \mathbb{R}^q, t' \in \mathbb{R}^q, \lambda \in [0,1]$. We have $V(t) = f(x(t),t), V(t') = f(x(t'),t')$ and

$$\begin{aligned} V(\lambda t + (1-\lambda)t') &\leq f(\lambda x(t) + (1-\lambda)x(t'), \lambda t + (1-\lambda)t') \\ &\leq \lambda f(x(t),t) + (1-\lambda)f(x(t'),t') = \lambda V(t) + (1-\lambda)V(t'). \end{aligned}$$

We have $V(t_0) = f(x(t_0),t_0)$ and $\forall t \in \mathbb{R}^q$, $V(t) = f(x(t),t) \leq f(x(t_0),t)$. From the previous proposition, V is differentiable at t_0 and $\nabla V(t_0) = f_2'(x(t_0),t_0)$. The proof is complete. □

Proposition 6.6.5. *Let f be a function from $\mathbb{R}^n$ into $\mathbb{R}$ and differentiable everywhere. Then f is convex if, and only if, $\forall x \in \mathbb{R}^n, \forall x_0 \in \mathbb{R}^n$, $f(x) - f(x_0) \geq \nabla f(x_0) \cdot (x - x_0)$.*

Proof. From Proposition 6.6.2, if f is convex then the inequality holds. Let us prove the converse. Let $x, y \in \mathbb{R}^n$. Let $t \in [0,1]$ and $z = tx + (1-t)y$. Observe that we have

$$f(z) = t(f(z) + \nabla f(z) \cdot (x-z)) + (1-t)(f(z) + \nabla f(z) \cdot (y-z)).$$

Assume the inequality is true. Then

$$f(z) \leq t(f(z)+f(x)-f(z))+(1-t)(f(z)+f(y)-f(z)) = tf(x)+(1-t)f(y).$$

□

6.7 Differential continuity for convex functions

In this section, we want to prove: (i) First, for a convex function, for $x \in$ i(dom(f)), if y is "close" to x then $\partial f(y)$ is "close" to $\partial f(x)$. An application of this result is that, if f is differentiable in i(dom(f)), then actually f is *continuously* differentiable in i(dom(f))
(ii) Second, if a sequence of convex functions $(f_i)_{i \in \mathbb{N}}$ from $\mathbb{R}^n$ into $\mathbb{R}$ converge pointwise to a function f (which is convex, continuous on $\mathbb{R}^n$, by Theorem 5.3.2), then for any $x \in \mathbb{R}^n$, when i is large enough, the subdifferentials $\partial f_i(x)$ will be "close" to $\partial f(x)$. Consequently, if the functions f_i, for $i \in \mathbb{N}$, and f are differentiable, then for any x, the gradients $\nabla f_i(x)$ converge to $\nabla f(x)$.
We first recall the following definition. Let $(a_i)_{i \in \mathbb{N}}$ be a sequence of real numbers, then by definition

$$\limsup_{i \to +\infty} a_i = \lim_{N \to +\infty} (\sup\{a_i \mid i \geq N\}).$$

This limit exists in $\overline{\mathbb{R}}$. Indeed, let $b_N = \sup\{a_i \mid i \geq N\}$. Then for all $N \in \mathbb{N}$, $b_N \in \mathbb{R} \cup \{+\infty\}$, and $\limsup_{i\to+\infty} a_i$ is the limit of this decreasing sequence (hence converging in $\overline{\mathbb{R}}$) $(b_N)_{N\in\mathbb{N}}$.

Lemma 6.7.1. *Let $(f_i)_{i\in\mathbb{N}}$ be a sequence of real-valued convex functions on a relatively open, convex set U, which satisfies $\limsup_{i\to+\infty} f_i(x) \leq f(x)$, $\forall x \in U$, where f is a real-valued, convex function on U. Then, for any compact subset K of U, for any $\varepsilon > 0$, there exists i_ε, such that $\forall i \geq i_\varepsilon, \forall x \in K$, $f_i(x) \leq f(x) + \varepsilon$.*

Proof. Define for every $x \in U$, for every $i \in \mathbb{N}$, $g_i(x) = \max\{f_i(x), f(x)\}$. The functions g_i are convex (Proposition 5.1.2, Assertion (i)) and real-valued. We want to prove that these functions converge pointwise to f.
First, for any $x \in U$, the sequence $(g_i(x))_{i\in\mathbb{N}}$ is bounded. Indeed, it is obvious that $\forall i \in \mathbb{N}$, $g_i(x) \geq f(x)$. Take some $\alpha > 0$. Since $\limsup_{i\to+\infty} f_i(x) \leq f(x)$, the set $\{i \in \mathbb{N} \mid f_i(x) > f(x) + \alpha\}$ is finite. Thus the sequence $(g_i(x))_{i\in\mathbb{N}}$ is bounded from above. Let $\ell(x)$ be an accumulation point of the sequence $(g_i(x))_{i\in\mathbb{N}}$. We have $\ell(x) = \lim_{k\to+\infty} g_{i_k}(x)$ for a subsequence $(g_{i_k}(x))_{k\in\mathbb{N}}$. By passing eventually to subsequences, we have either $g_{i_k}(x) = f_{i_k}(x) \geq f(x), \forall k \in \mathbb{N}$ or $g_{i_k}(x) = f(x), \forall k \in \mathbb{N}$. In the first case, we have $g_{i_k}(x) \leq \sup\{f_i(x) \mid i \geq i_k\}$ and hence $f(x) \leq \ell(x) \leq \lim_{k\to+\infty} \sup\{f_i(x) \mid i \geq i_k\} = \limsup_{i\to+\infty} f_i(x) \leq f(x)$. In both cases, we have $\ell(x) = f(x)$. We have proved that the sequence $(g_i(x))_{i\in\mathbb{N}}$ converges to $f(x)$.
It follows from Theorem 5.3.2 that the sequence $(g_i)_{i\in\mathbb{N}}$ converges uniformly to f on any compact set K of U. The statement is therefore true. □

Proposition 6.7.1. *Let f be a real-valued, convex function on a convex, open subset U of $\mathbb{R}^n$. Let $(f_i)_{i\in\mathbb{N}}$ be a sequence of real-valued convex functions on U which converges pointwise to f. Consider a sequence $(x_i)_{i\in\mathbb{N}}$ of points in U which converges to a point $x \in U$. Then, for any $\varepsilon > 0$, there exists i_ε such that*

$$\forall i \geq i_\varepsilon,\ \partial f_i(x_i) \subset \partial f(x) + \varepsilon\overline{B(0,1)}.$$

Proof. Let $\alpha > 0$. Define $\mu = f'(x;y) + \alpha$, where $y \in \mathbb{R}^n$. There exists a $\lambda > 0$ such that $x + \lambda y \in \mathrm{i}(\mathrm{dom}(f))$ and $(f(x+\lambda y) - f(x))/\lambda < \mu$. By Theorem 5.3.2, $f_i(x_i + \lambda y)$ tends to $f(x + \lambda y)$ and $f_i(x_i)$ tends to $f(x)$. Hence, there exists i_1 such that $\forall i \geq i_1$, $(f_i(x_i + \lambda y) - f_i(x_i))/\lambda < \mu$. Since $f_i'(x_i;y) \leq (f_i(x_i + \lambda y) - f_i(x_i))/\lambda$, we have $\forall i \geq i_1$, $f_i'(x_i;y) \leq \mu$, which implies $\limsup_{i\to+\infty} f_i'(x_i;y) \leq \mu = f'(x;y) + \alpha$. Since this inequality holds for any $\alpha > 0$, we actually have $\limsup_{i\to+\infty} f_i'(x_i;y) \leq f'(x;y)$. We apply Lemma 6.7.1 to have

$$\forall \varepsilon > 0, \exists i_\varepsilon \in \mathbb{N}, \forall i \geq i_\varepsilon, \forall y \in \overline{B(0,1)},\ f_i'(x_i;y) \leq f'(x;y) + \varepsilon.$$

Now, if $y \in \mathbb{R}^n, y \neq 0$, from Assertion (ii) of Proposition 6.4.1,

$$(1/\|y\|)f'(x_i;y) = f'(x_i;y/\|y\|) \leq f'(x;y/\|y\|) + \varepsilon = f'(x;y)/\|y\| + \varepsilon.$$

Hence $\forall y \in \mathbb{R}^n$, $f'(x_i; y) \leq f'(x; y) + \varepsilon\|y\|$. Denote by g the function $g(y) = \|y\|$, $\forall y \in \mathbb{R}^n$. This function is obviously real-valued and sublinear. We have $\Delta(f'(x_i; .)) \subset \Delta(f'(x; .) + \varepsilon g)$ (Proposition 6.2.2).
But $\Delta(f'(x_i; .)) = \partial f(x_i)$ (Proposition 6.5.1), $\Delta(f'(x; .) + \varepsilon g) = \Delta(f'(x; .)) + \varepsilon\Delta(g)$ (Proposition 6.3.5), and $\Delta(g) = \overline{B(0,1)}$ (Exercise 6.8). The proof is complete. □

A particular case of the previous proposition is the following:

Proposition 6.7.2. *Let f be a convex function on $\mathbb{R}^n$. Let $x \in \mathrm{i}(\mathrm{dom}(f))$ and $(x_i)_{i\in\mathbb{N}}$ be a sequence of points in $\mathrm{i}(\mathrm{dom}(f))$ converging to x. Then, for any $\varepsilon > 0$, there exists i_ε in $\mathbb{N}$ such that $\forall i \geq i_\varepsilon$, $\partial f(x_i) \subset \partial f(x) + \varepsilon\overline{B(0,1)}$.*

Proof. Apply Proposition 6.7.1 with $U = \mathrm{i}(\mathrm{dom}(f))$ and $f_i = f$, $\forall i$. □

As corollary of Proposition 6.7.2, we have

Corollary 6.7.1. *Let f be a convex function on $\mathbb{R}^n$, $(x_i)_{i\in\mathbb{N}}$ be a sequence of points of $\mathrm{i}(\mathrm{dom}(f))$ converging to $x \in \mathrm{i}(\mathrm{dom}(f))$, $(x_i^*)_{i\in\mathbb{N}}$ be a sequence of elements of $\mathbb{R}^n$ which satisfy $\forall i \in \mathbb{N}$, $x_i^* \in \partial f(x_i)$. Then there exists a subsequence of $(x_i^*)_{i\in\mathbb{N}}$ which converges to some $x^* \in \partial f(x)$.*

Proof. From the previous proposition, given any $\varepsilon > 0$, there exists i_ε in $\mathbb{N}$ such that $\forall i \geq i_\varepsilon$, $\partial f(x_i) \subset \partial f(x) + \varepsilon\overline{B(0,1)}$. Since $x \in \mathrm{i}(\mathrm{dom}(f))$, $\partial f(x)$ is compact (Proposition 6.5.2, Assertion (iii)) and hence $\partial f(x) + \varepsilon\overline{B(0,1)}$ is also compact. The sequence $(x_i^*)_{i\in\mathbb{N}}$ for $i \geq i_\varepsilon$ will be in a compact set. There exists a subsequence converging to some x^*.
Let $z \in \mathbb{R}^n$. We have $\forall i \in \mathbb{N}$, $f(z) - f(x_i) \geq {x_i}^* \cdot (z - x_i)$. Since f is continuous in $\mathrm{i}(\mathrm{dom}(f))$ (Theorem 5.2.1), we get, by taking the limits when i tends to $+\infty$, that $f(z) - f(x) \geq x^* \cdot (z - x)$. Since z is arbitrary in $\mathbb{R}^n$, we have $x^* \in \partial f(x)$. □

Another interesting corollary of Proposition 6.7.2 is the following

Proposition 6.7.3. *Let f be a convex function on $\mathbb{R}^n$. If f is differentiable in $\mathrm{i}(\mathrm{dom}(f))$, then actually f is continuously differentiable in $\mathrm{i}(\mathrm{dom}(f))$.*

Proof. Let $x \in \mathrm{i}(\mathrm{dom}(f))$ and let x_i be a sequence of points of $\mathrm{i}(\mathrm{dom}(f))$ which converge to x. From the previous proposition, for any $\varepsilon > 0$ there exists i_ε such that $\forall i \geq i_\varepsilon$, $\|\nabla f(x_i) - \nabla f(x)\| \leq \varepsilon$. □

If f is a real-valued function defined on an interval I of $\mathbb{R}$, we will denote for all x in I, $f'_+(x) = f'(x; 1)$ and $f'_-(x) = f'(x; -1)$. Recall that in real analysis, $f'_+(x)$ and $f'_-(x)$ denote the right and the left derivatives at the point x of the function f.

Proposition 6.7.4. *Let f be a convex function from $[a,b]$ into $\mathbb{R}$. Then: (i) f has real-valued left and right derivatives, f'_-, f'_+, in $]a,b[$. In fact, we have $\forall x \in]a,b[\,,\ \partial f(x) = [f'_-(x), f'_+(x)]$. If $a < x < x' < b$ then*

$$f'_-(x) \leq f'_+(x) \leq \frac{f(x')-f(x)}{x'-x} \leq f'_-(x') \leq f'_+(x').$$

If f is differentiable in $]a,b[$, then it is continuously differentiable in $]a,b[$. (ii) Both $f'_+(a)$ and $f'_-(b)$ exists in $\overline{\mathbb{R}}$ ($f'_+(a)$ may be equal to $-\infty$, $f'_-(b)$ may be equal to $+\infty$). If f is continuous at a (respectively b), and if f is differentiable in $]a,b[$, then $f'(x)$ converges to $f'_+(a)$ (resp. $f'_-(b)$) when x tends to a (resp. b).

Proof. (i) Let $x \in]a,b[$. From Proposition 6.5.2, $\partial f(x) \neq \emptyset$ and $\partial f(x) = [\alpha,\beta]$. We have $f'_+(x) = \lim_{y>0, y\to 0}(f(x+y\cdot 1) - f(x))/y = f'(x;1)$ that is finite since $x \in \mathrm{i}(\mathrm{dom}(f))$. Since $f(x+y) - f(x) \geq \beta y, \forall y \in \mathbb{R}$, we have $f'(x;1) \geq \beta$. Now, if $y \geq 0$, one has $f'(x;1)y = f'(x;y)$ (Assertion (ii) of Proposition 6.4.1), and if $y < 0$ then $f'(x;1)y = -f'(x;1)(-y) = -f'(x;-y) \leq f'(x;y)$ (Proposition 6.4.1). That means that $f'(x;1) \in \partial f(x)$ (Proposition 6.5.1). Hence $f'(x;1) \leq \beta$ which implies $f'_+(x) = \beta$. Similarly, one proves that $f'_-(x) = \alpha$. We have

$$f'_+(x) = f'(x;1) = \inf_{y>0} \frac{f(x+y)-f(x)}{y} \leq \frac{f(x')-f(x)}{x'-x} \text{ if } x' > x.$$

Similarly,

$$-f'_-(x') = \inf_{y>0} \frac{f(x+y\cdot(-1)) - f(x))}{y} \leq \frac{f(x)-f(x')}{x'-x}.$$

Hence the required inequality holds.
Assume now that f is differentiable in $]a,b[$. One can apply the previous proposition to state that f is continuously differentiable in $]a,b[$. We give here an elementary proof. We have seen that the derivative is a nondecreasing function. Assume that y tends to x with $y > x$. Then $f'(y)$ converges to some ℓ. We have $\forall z \in \mathbb{R}^n$, $f(z) - f(y) \geq f'(y)(z-y)$. Hence $\forall z \in \mathbb{R}^n$, $f(z) - f(x) \geq \ell(z-x)$. That means that $\ell \in \partial f(x)$. Hence $\ell = f'(x)$ from Proposition 6.6.1. We have proved that f is continuously differentiable in $]a,b[$.
(ii) Since $f'_+(a) = f'(a,1)$ (resp. $f'_-(b) = f'(b,-1)$), it is the limit of $(f(y) - f(a))/(y-a)$ (resp. $(f(y)-f(b))/(y-b)$) when y tends to a (resp. to b).
We will prove the second claim. First assume that $f'_+(a) > -\infty$. Since $f'(y) \geq f'_+(a), \forall y \in]a,b[$, and since $f'(y)$ decreases when y tends to a, then $f'(y)$ converges to $\ell \geq f'_+(a)$. We can show as before that $\forall z \in \mathbb{R}^n$, $f(z) - f(a) \geq \ell(z-a)$ which implies $\ell \leq f'_+(a)$. Hence $\ell = f'_+(a)$. Similarly, if $f'_-(b) < +\infty$ then $f'(y)$ converges to $f'_-(b)$ when y tends to b.
Assume now that $f'_+(a) = -\infty$. We will show that $f'(y)$ converges to $-\infty$

when y tends to a. Since $f'(y)$ decreases when y tends to a, it will have a limit ℓ. If ℓ is different from $-\infty$, one can see as before that $\ell \in \partial f(a)$. In particular, $\ell \leq (f(y) - f(a))/(y - a), \forall y > a$. But this ratio converges to $-\infty$ which contradicts that $\ell > -\infty$. We have the same conclusion when $f'_-(b) = +\infty$. □

Corollary 6.7.2. *Let f be a convex function from $]a, b[$ (a may be equal to $-\infty$, b may be $+\infty$) into $\mathbb{R}$.*
(i) If $\forall x \in]a, b[$, $f'_-(x) \leq 0$ (respectively $f'_+(x) \geq 0$), then f is nonincreasing (resp. nondecreasing) on $]a, b[$.
(ii) If f is nonincreasing (respectively nondecreasing) on $]a, b[$ then $\forall x \in]a, b[$, $f'_+(x) \leq 0$ (respectively $f'_-(x) \geq 0$).
(iii) If $f_+(x) = 0$, $\forall x \in]a, b[$ or $f_-(x) = 0$, $\forall x \in]a, b[$, then f is a constant function.

Proof. Assertions (i), (ii) and (iii) follow from the inequality stated in Proposition 6.7.4: if $a < x < x' < b$ then

$$f'_-(x) \leq f'_+(x) \leq \frac{f(x') - f(x)}{x' - x} \leq f'_-(x') \leq f'_+(x').$$ □

The following proposition is another particular case of Proposition 6.7.1

Proposition 6.7.5. *Let $(f_i)_{i \in \mathbb{N}}$ be a sequence of convex functions from $\mathbb{R}^n$ into $\mathbb{R}$ which converge pointwise to a function f (which is convex from $\mathbb{R}^n$ into $\mathbb{R}$, by Theorem 5.3.2). Let $x \in \mathbb{R}^n$.*
(i) Then for any $\varepsilon > 0$, there exists i_ε in $\mathbb{N}$ such that $\forall i \geq i_\varepsilon$, $\partial f_i(x) \subset \partial f(x) + \varepsilon B(0, 1)$.
(ii) If the functions f_i, for $i \in \mathbb{N}$, and f are differentiable at x, then the gradients $\nabla f_i(x)$ converge to $\nabla f(x)$.

Proof. (i) Apply Proposition 6.7.1 with $U = \mathbb{R}^n$ and $x_i = x$, $\forall i \in \mathbb{N}$.
(ii) Obviously, when the functions f_i, for $i \in \mathbb{N}$, and f are differentiable at x, then the gradients $\nabla f_i(x)$ converge to $\nabla f(x)$, by (i). □

Exercises

Exercise 6.1. Demonstrate Proposition 6.2.1.

Exercise 6.2. Demonstrate Corollary 6.2.1, Propositions 6.3.1, 6.3.2.

Exercise 6.3. Demonstrate Assertions (ii), (iii) of Proposition 6.4.1.

Exercise 6.4. Demonstrate the statements of Remark 6.3.1.
Hint: Use Corollary 2.1.4

Exercise 6.5. Let f be the function as follows. $f(x) = (x+1)^2$ if $x \geq 0$; $f(x) = (x-1)^2$, if $x \leq 0$.
a) Show that f is convex.
b) Determine $\partial f(0)$ and $\partial f(x)$ for $x \neq 0$.

Exercise 6.6. Let f be a convex function from $\mathbb{R}^n$ into $\mathbb{R} \cup \{+\infty\}$. Show that $\overline{x}$ minimizes f if, and only if $0 \in \partial f(\overline{x})$. Let $g : [0,1] \to \mathbb{R} \cup \{+\infty\}$ be defined by $g(x) = 1 - x$ if $x \in [0,1]$, and by $g(x) = +\infty$ if $x \notin [0,1]$.
a) Show that g is convex.
b) For any $x \in \mathbb{R}$, determine $\partial g(x)$.
c) Find the point which minimizes g.

Exercise 6.7. Let f be a function from $\mathbb{R}^n$ into $\mathbb{R}$ such that $\forall x \in \mathbb{R}^n, \forall y \in \mathbb{R}^n$, $\lim_{\lambda>0,\lambda\to 0}\big(f(x+\lambda y) - f(x)\big)/\lambda$ exists. We denote this limit by $f'(x;y)$. We assume that for all x in $\mathbb{R}^n$, the function $\mathbb{R}^n \to \mathbb{R}$ defined by $y \to f'(x;y) \in \mathbb{R}$ is convex. Show that f is convex if, and only if, $f(y) - f(x) \geq f'(x; y-x)$, $\forall x \in \mathbb{R}^n, \forall y \in \mathbb{R}^n$.

Exercise 6.8. For $x = (x_1, \ldots, x_n) \in \mathbb{R}^n$ define $f(x) = \sqrt{x_1^2 + \ldots + x_n^2}$. Show that f is subdifferentiable everywhere and $\partial f(0) = \{x \in \mathbb{R}^n \mid \sqrt{x_1^2 + \ldots + x_n^2} \leq 1\} = \overline{B(0,1)}$ and hence $\delta^*(\overline{B(0,1)})(y) = \|y\|, \forall y \in \mathbb{R}^n$.

Exercise 6.9. Let $(e^1, \ldots, e^n)$ denote the canonical basis of $\mathbb{R}^n$. Show that: $\mathrm{co}\{e^1, \ldots, e^n, -e^1, \ldots, -e^n\} = \{(\sum \lambda_i e^i) \mid \sum |\lambda_i| \leq 1\}$. Show also that $\partial f(0) = \mathrm{co}\{e^1, \ldots, e^n, -e^1, \ldots, -e^n\}$, where $f(x) = \max_i |x_i|$.
Let $(a_1, \ldots, a_n) \in \mathbb{R}^n$. Find the maximum of $\sum_i a_i x_i$ where $(x_1, \ldots, x_n) \in \mathrm{co}\{e^1, \ldots, e^n, -e^1, \ldots, -e^n\}$.

Exercise 6.10. Let p, q be two positive rational numbers such that $1/p + 1/q = 1$. Let $f(x) = (|x_1|^p + \ldots, +|x_n|^p)^{\frac{1}{p}}$, $g(x) = (|x_1|^q + \ldots, +|x_n|^q)^{\frac{1}{q}}$, where $x = (x_1, \ldots, x_n) \in \mathbb{R}^n$. We know that f and g are convex. Show that $\partial f(0) = \{x \in \mathbb{R}^n \mid g(x) \leq 1\}$.

Exercise 6.11. Let $f : \mathbb{R}^n \to \mathbb{R} \cup \{+\infty\}$ be defined by $f(x) = -\sqrt{1 - \|x\|^2}$ if $\|x\| \leq 1$, and by $f(x) = +\infty$ if $\|x\| > 1$.
a) Show that f is convex.
b) Show that $\partial f(x) = \emptyset$ if $\|x\| \geq 1$, and $\partial f(x) \neq \emptyset$ if $\|x\| < 1$.

Exercise 6.12. Let f, g be two sublinear functions from $\mathbb{R}^n$ into $\mathbb{R}$. Let $h = \sup(f, g)$, i.e. $\forall x \in \mathbb{R}^n$, $h(x) = \max(f(x), g(x))$.
Show that $\Delta(h) = \mathrm{co}\{\Delta(f), \Delta(g)\}$. Hence, if $f_1, \ldots, f_p$ are sublinear functions from $\mathbb{R}^n$ into $\mathbb{R}$ and if $h = \sup_i f_i$, then $\Delta(h) = \mathrm{co}\{\Delta(f_1), \ldots, \Delta(f_p)\}$.
Applications
a) Let $f(x) = \max_i |x_i|$ where $x = (x_1, \ldots, x_n)$. Show that $\partial f(0) = \mathrm{co}\{e^1, \ldots, e^n, -e^1, \ldots, -e^n\}$, where $(e^1, \ldots, e^n)$ is the canonic basis of $\mathbb{R}^n$.
b) Let $f_1, \ldots, f_p$ be p convex functions from $\mathbb{R}^n$ into $\mathbb{R}$ and let $f = \sup_i f_i$. Let $I(x) = \{i \in \{1, \ldots, p\} \mid f(x) = f_i(x)\}$. Show that $\partial f(x_0) = \mathrm{co}\{\partial f_i(x_0), i \in I(x_0)\}$.

Exercise 6.13. a) Let $f :]a, +\infty[\to \mathbb{R}$ be a convex function. Show that if f is bounded from above, then f is nonincreasing.
b) Let $f : \mathbb{R} \to \mathbb{R}$ be a convex function. Show that if f is bounded from above, then f is a constant function.

Exercise 6.14. Let f be a real-valued function defined on a relatively open convex subset A of $\mathbb{R}^n$. The function f is said to be subdifferentiable at $x \in A$ if there exists $x^* \in \mathbb{R}^n$ such that $f(z) \geq f(x) + x^* \cdot (z - x)$, $\forall z \in A$. Prove that f is convex if f is subdifferentiable at each point of A.

Exercise 6.15 (Cones, orderings and extension of positive linear functionals). A convex cone P such that $P \cap (-P) = \{0\}$ defines a reflexive, transitive and anti-symmetric partial ordering on $\mathbb{R}^n$ by

$$x \leq y \Longleftrightarrow y - x \in P.$$

This ordering has the following properties that for any $z \in \mathbb{R}^n$ and $\lambda \in \mathbb{R}_+$, $x \leq y \Rightarrow x + z \leq y + z$ and $\lambda x \leq \lambda y$. The cone P is said to be the *positive cone* of this linear ordering and $\mathbb{R}^n$ is said to be *(linearly) ordered* by P. Every linear subspace M of $\mathbb{R}^n$ is then (linearly) ordered by $P \cap M$. If $\mathbb{R}^n$ is ordered by P, a linear functional f on $\mathbb{R}^n$ is said to be *positive* if $f(x) \geq 0$ whenever $x \in P$. The set P^+ denotes the cone of positive linear functionals on $\mathbb{R}^n$. Identifying a linear functional f with the vector $p \in \mathbb{R}^n$ such that $f(x) = p \cdot x$, $\forall x \in \mathbb{R}^n$, one obviously has $P^+ = -P^\circ$.
a) Show that if $\mathrm{i}(P) \neq \emptyset$ then $\mathbb{R}^n = P - P$ and deduce that any two elements of $\mathbb{R}^n$ have then an upper bound (i.e. that given $x, y \in \mathbb{R}^n$, there exists $z \in \mathbb{R}^n$ such that $x \leq z$ and $y \leq z$).
b) Let M be a linear subspace of $\mathbb{R}^n$ such that $M \cap \mathrm{i}(P) \neq \emptyset$. The purpose of this question is to deduce from the Hahn-Banach dominated extension theorem that every positive linear functional h on M admits a positive extension to all of $\mathbb{R}^n$.
(i) Show that for every $x \in \mathbb{R}^n$ there exists $y \in M$ such that $y - x \in P$.
(ii) Let $g : \mathbb{R}^n \to \overline{\mathbb{R}}$ be the function defined by

$$g(x) := \inf\{h(y) \mid y \in M \text{ and } x \leq y\}.$$

Show that g is a real-valued sublinear function such that $h(x) \leq g(x)$, $\forall x \in M$.
(iii) In view of Theorem 6.1.1, let $h_1 : \mathbb{R}^n \to \mathbb{R}$ be a linear functional extending h. Show that $h_1(x) \geq 0$, $\forall x \in P$.
Hint: Consider $x \in P$, $y + \lambda x$ for $y \in P \cap M$ and show that $h_1(-x) \leq 0$.
c) In $\mathbb{R}^3$, consider the cone C generated by the set $\{(x_1, x_2, x_3) \in \mathbb{R}^3 \mid x_3 = 1, x_2 \geq (x_1^3)^-\}$ and the hyperplane M of equation $x_2 = 0$. Give an example of positive linear functional on M which does not admit any positive extension to $\mathbb{R}^3$.

Exercise 6.16. A convex function $p : \mathbb{R}^n \to \mathbb{R}$ such that $p(\lambda x) = |\lambda| p(x)$ for all $x \in \mathbb{R}^n$ and scalar λ is called a *semi-norm* on $\mathbb{R}^n$. Clearly, a semi-norm p on $\mathbb{R}^n$ is positive ($x \geq 0$ implies $p(x) \geq 0$) and $p(0) = 0$. Using the subdifferentiability theorem, show that given a semi-norm p on $\mathbb{R}^n$, for each $x_0 \in \mathbb{R}^n$ there exists a linear functional on $\mathbb{R}^n$ such that

$f(x_0) = p(x_0)$ and

$|f(x)| \leq p(x)$ for all $x \in \mathbb{R}^n$.

Exercise 6.17. The purpose of this exercise is to obtain the separation theorems from the subdifferentiability theorem. Actually, it suffices to show that Proposition 2.1.6 can be obtained from this theorem. It will be done in several steps.

Let C be a closed convex subset of $\mathbb{R}^n$ and $x_0 \notin C$. Assume $0 \in \mathrm{ri}(C)$.

a) Show that there exists a closed, convex subset D of F such that $C \subset \mathrm{i}(D)$ and $x_0 \notin D$.

Hint: Show that there exists $\varepsilon > 0$ such that $x_0 \notin \overline{C + \varepsilon B(0,1)}$.

b) Let γ_D denote the gauge of D, i.e. the function defined on $\mathbb{R}^n$ by $\gamma_D(x) = \inf\{\lambda > 0 \mid (1/\lambda)x \in D\}$.

(i) Show that $\gamma_D(x) \in \mathbb{R}_+, \forall x \in \mathbb{R}^n$, γ_D is sublinear, $\gamma_D(x_0) > 1$, $\gamma_D(x) < 1$, if $x \in \mathrm{i}(D)$. (See Exercise 1.16.)

(ii) Noticing that $\partial\gamma_D(x_0) \neq \emptyset$, take $p \in \partial\gamma_D(x_0)$. Show that $p \cdot x \leq \alpha < p \cdot x_0$, $\forall x \in C$, where $\alpha = p \cdot x_0 + 1 - \gamma_D(x_0)$.

7. Convex Optimization With Convex Constraints

In this chapter we want to solve the problem $\min\{f(x) \mid x \in C\}$, where f is a convex function on $\mathbb{R}^n$, and C is a convex, nonempty subset of $\mathbb{R}^n$. A point $x^* \in C$ is a *global solution*, or more simply a *solution* to this problem, or a *minimizer* of f on C, if $f(x^*) \leq f(x)$, $\forall x \in C$. We say that x^* is a *local solution* to this problem if there exists a relatively open set U of C such that $f(x^*) \leq f(x)$, $\forall x \in U$.

In the first section of this chapter we will give conditions under which the minimization problem of a convex function f has a solution. The idea is to give conditions under which the problem turns out to be a minimization problem on a compact set. Since we will assume, in this section, that f is real-valued and hence continuous, the problem has a solution by a well- known argument in Topology: a continuous function always has a minimum on any compact set. But in fact one just needs, as in Rockafellar ([51]), to assume that f is lower-semicontinuous (i.e. any level set $\text{lev}_{\leq\alpha}(f)$ is closed) and takes values in $\mathbb{R} \cup \{+\infty\}$, because any lower-semicontinuous function has an attained minimum on any compact set. For simplicity, in the first section, we assume that this function f is continuous.

In the second section, the set C of constraints will be defined by a finite number of constraints $f_i(x) \leq 0$, $\forall i \in I$. We give necessary and sufficient criteria to check that a point is a minimizer of a convex function. These criteria are known as *Kuhn-Tucker Conditions.*

But we can define the set C_α by the constraints $f_i(x) \leq \alpha_i$, $\forall i \in I$. The minimum of the function f on C_α is therefore a function of the parameters α_i. This function is called *value function* of the minimization problem, or more simply, the value function. An interesting question is how to measure the marginal variation of the minimum when the parameters α_i change. In the third section we will show that the value function is subdifferentiable with respect to the parameters α_i. So, the values of the changes in the minimum when the parameters vary can be "measured" by the values of the subdifferential.

Throughout this chapter, we will adopt the following rules which have been already mentioned in Chapters 5 and 6. For any convex function f, if $\lambda = 0$, the function (λf) is the zero function, and $\lambda\partial f(x) = \{0\}$. With these rules, we always have $\partial(\lambda f)(x) = \lambda\partial f(x)$, $\forall\lambda \geq 0$.

7.1 The minimum of a convex function $f : \mathbb{R}^n \longrightarrow \mathbb{R}$

In this section, we want to find the infimum under convex constraints of a real-valued convex function on $\mathbb{R}^n$. We first consider the infimum over $\mathbb{R}^n$ of a convex function.

Let us recall the definition of the level sets of a convex function f. For $\alpha \in \mathbb{R}$, $\text{lev}_{\leq\alpha}(f) = \{x \in \mathbb{R}^n \mid f(x) \leq \alpha\}$. We know that $\text{lev}_{\leq\alpha}(f)$ is convex for any α (see Exercise 5.6). Since f is real-valued, f is continuous over $\mathbb{R}^n$ (Corollary 5.2.2), and hence $\text{lev}_{\leq\alpha}(f)$ is closed for any α. The infimum over $\mathbb{R}^n$ of f is $\inf f = \inf\{\alpha \mid \text{lev}_{\leq\alpha}(f) \neq \emptyset\}$.

Let us now introduce the asymptotic directions of a real-valued convex function f. We need the following result for the asymptotic cones of the sets $\text{lev}_{\leq\alpha} f$.

Proposition 7.1.1. *Let f be a convex, real-valued function on $\mathbb{R}^n$. Then the nonempty levels sets $\text{lev}_{\leq\alpha}(f)$ have the same asymptotic cone.*

Proof. Let α_1, α_2 satisfy $\alpha_1 < \alpha_2, \text{lev}_{\leq\alpha_1}(f) \neq \emptyset, \text{lev}_{\leq\alpha_2}(f) \neq \emptyset$. Let A_1 and A_2 denote the asymptotic cones of these level sets. Recalling that these level sets are closed, since $\text{lev}_{\leq\alpha_1} f \subset \text{lev}_{\leq\alpha_2} f$, we know that $A_1 \subset A_2$ (see Exercise 1.17, Assertion (vi)).

Let us prove the converse. Let $y \in A_2$. Take $x \in \text{lev}_{\leq\alpha_1}(f) \subset \text{lev}_{\leq\alpha_2}(f)$. We have $\forall\lambda > 0$, $f(x+\lambda y) \leq \alpha_2$. From Exercise 6.13, the function $g : \mathbb{R}_+ \setminus \{0\} \to \mathbb{R}$ defined by $\lambda \to f(x+\lambda y)$ is nonincreasing. Let $(\lambda_i)_{i\in\mathbb{N}}$ be a sequence of positive numbers tending to zero. We have, by continuity of f, that $(g(\lambda_i))_{i\in\mathbb{N}} = (f(x+\lambda_i y))_{i\in\mathbb{N}}$ converges to $f(x)$ when i tends to $+\infty$. Hence $\forall\lambda > 0$, there exists N_λ such that for any $i \geq N_\lambda$, $f(x+\lambda y) \leq f(x+\lambda_i y)$ and by passing to limits, $f(x+\lambda y) \leq f(x) \leq \alpha_1$. Thus $x+\lambda y \in \text{lev}_{\leq\alpha_1}(f)$. Since λ is arbitrary in $\mathbb{R}_+ \setminus \{0\}$, we have $y \in A_1$. □

The previous proposition allows us to give the following definition.

An *asymptotic direction of a function* f is an asymptotic direction of a nonempty level set $\text{lev}_{\leq\alpha}(f)$. We say that f has no asymptotic direction if the asymptotic cone of all the nonempty level sets of f is equal to $\{0\}$.

Proposition 7.1.2. *Let f be a real-valued, convex function on $\mathbb{R}^n$ which has no asymptotic direction. Then the infimum of f is finite and attained.*

Proof. Since f is continuous, it suffices to show that any nonempty level set of f is compact. By assumption, f has no asymptotic direction or equivalently, the common asymptotic cone of the nonempty level sets of f is equal to $\{0\}$. By Exercise 1.17, every level set is bounded. Since the level sets are closed, the proof is complete. □

Proposition 7.1.3. *Let f be a real-valued, convex function on $\mathbb{R}^n$ which has no asymptotic direction. Let $(x_i)_{i\in\mathbb{N}}$ be a sequence of points of $\mathbb{R}^n$ such that $\lim_{i\to+\infty} f(x_i) = \inf f$. Then*

(i) The sequence $(x_i)_{i\in\mathbb{N}}$ is bounded. Any accumulation point of this sequence is a minimizer of f.
(ii) If f has a unique minimizer x then the sequence $(x_i)_{i\in\mathbb{N}}$ converges to x.

Proof. (i) Take some $\varepsilon > 0$ and let $\alpha = \inf f + \varepsilon$. For any i large enough, $f(x_i) \le \alpha$, or equivalently $x_i \in \text{lev}_{\le\alpha}(f)$. But, by assumption, f has no asymptotic direction and consequently, $\text{lev}_{\le\alpha}$ is bounded. Hence, the sequence $(x_i)_{i\in\mathbb{N}}$ is bounded.
If x^* is an accumulation point of this sequence then $x^* = \lim_{k\to+\infty} x_{i_k}$ for a subsequence $(x_{i_k})_{k\in\mathbb{N}}$. By continuity, $f(x^*) = \lim_{k\to+\infty} f(x_{i_k}) = \lim_{i\to+\infty} f(x_i) = \inf f$.

(ii) From (i), the sequence is bounded and this minimizer is a unique accumulation point. Therefore the whole sequence $(x_i)_{i\in\mathbb{N}}$ converges to this point. □

Remark 7.1.1. Note that if f has a unique minimizer then f has no asymptotic direction because the level set $\text{lev}_{\le\inf f}(f)$ is a singleton, hence compact.

Proposition 7.1.4. *Let f and $\{f_i\}_{i\in I}$ be real-valued, convex functions on $\mathbb{R}^n$, where I is an arbitrary index set. Assume there exists $x_0 \in \mathbb{R}^n$ which satisfy $f_i(x_0) \le 0, \forall i \in I$. Assume also that there is no common asymptotic direction for the function f and all the functions $(f_i)_{i\in I}$. Then the infimum of f under the constraints $f_i(x) \le 0$, $\forall i \in I$ is attained.*

Proof. Let $C = \{x \in \mathbb{R}^n \mid f_i(x) \le 0,\ \forall i \in I\}$. The set C is convex and closed. It is nonempty by assumption. Let $A = \{x \in \mathbb{R}^n \mid x \in C \text{ and } f(x) \le f(x_0)\}$. Note that A is convex, closed, nonempty ($x_0 \in A$). An asymptotic direction of A is a common asymptotic direction of f and all of the $\{f_i\}_{i\in I}$. By assumption, this common asymptotic direction is $\{0\}$. Hence A is bounded (Exercise 1.17). □

7.2 Kuhn-Tucker Conditions

7.2.1 Necessary and sufficient condition for optimality

The aim of this section is to give the necessary and sufficient conditions for a point to be an optimal solution to Problem (P):

$$(P) \qquad \begin{array}{l} \text{Minimize } f_0(x) \text{ under the constraints} \\ \begin{cases} f_i(x) \le 0, & \forall i \in I \\ g_i(x) \le 0, & \forall i \in J \\ g_i(x) = 0, & \forall i \in K. \end{cases} \end{array}$$

where $f_0 : \mathbb{R} \to \mathbb{R}^n \cup \{+\infty\}$ is a convex function, I, J and K are finite and possibly empty sets, for all $i \in I$, f_i is convex, non-affine function from $\mathbb{R}^n$

into $\mathbb{R}\cup\{+\infty\}$, for all $i \in J\cup K$, g_i is a non-null affine function. We will also assume that $\operatorname{ri}(\operatorname{dom}(f_0)) \cap (\cap_{i\in I} \operatorname{ri}(\operatorname{dom}(f_i))) \neq \emptyset$.

The function f_0 is called the *objective function.* A *feasible point* is a point $x \in \mathbb{R}^n$ that satisfies all the constraints. An *optimal solution* to (P) or simply a *solution* to (P) is a feasible point $\overline{x}$, such that for all feasible point x, $f_0(x) \geq f_0(\overline{x})$. The *domain of Problem* (P) is the set $\operatorname{dom}(P) := \operatorname{dom}(f_0) \cap (\cap_{i\in I} \operatorname{dom}(f_i))$. Notice that if the set I is empty, then $\operatorname{dom}(P) = \operatorname{dom}(f_0)$. Notice also that we have $\operatorname{ri}(\operatorname{dom}(P)) = \operatorname{ri}(\operatorname{dom}(f_0)) \cap (\cap_{i\in I} \operatorname{ri}(\operatorname{dom}(f_i))) \neq \emptyset$ (see Exercise 1.26).

Rockafellar ([51]) assumes that $\forall i \in I$, $\operatorname{ri}(\operatorname{dom}(f_0)) \subset \operatorname{ri}(\operatorname{dom}(f_i))$, and $\operatorname{dom}(f_0) \subset \operatorname{dom}(f_i)$. In this case we have $\operatorname{dom}(P) = \operatorname{dom}(f_0)$ and $\operatorname{ri}(\operatorname{dom}(f_0)) \cap (\cap_{i\in I} \operatorname{ri}(\operatorname{dom}(f_i))) = \operatorname{ri}(\operatorname{dom}(f_0))$ and hence is nonempty.

Remark 7.2.1. Let us recall that for a convex function f from $\mathbb{R}^n$ into $\mathbb{R}\cup\{+\infty\}$, $\overline{x}$ minimizes f on $\mathbb{R}^n$ if, and only if, $0 \in \partial f(\overline{x})$ (see Exercise 6.6).

Lemma 7.2.1. *Let C be a nonempty convex set of $\mathbb{R}^n$. Let I, J and K be finite possibly empty sets, and for all $i \in I$, f_i is a convex, non-affine function from C into $\mathbb{R}$, and for all $i \in J \cup K$, g_i is non-null affine function. Assume there exists $x_0 \in \operatorname{ri}(C)$ such that*

$$\begin{cases} g_i(x_0) \leq 0, & \forall i \in J \\ g_i(x_0) = 0, & \forall i \in K. \end{cases}$$

If the system:

$$\begin{cases} f_i(x) < 0, & \forall i \in I \\ g_i(x) \leq 0, & \forall i \in J \\ g_i(x) = 0, & \forall i \in K. \end{cases}$$

has no solution in C, then there exist nonnegative real scalars $(\lambda_i)_{i\in I}$, $(\mu_i)_{i\in J}$, and real numbers $(\mu_i)_{i\in K}$, at least one of the $(\lambda_i)_{i\in I}$ is not zero, which verify:

$$\sum_{i\in I} \lambda_i f_i(x) + \sum_{i\in J} \mu_i g_i(x) + \sum_{i\in K} \mu_i g_i(x) \geq 0, \ \forall x \in C.$$

Proof. Let $p = \operatorname{card}(I)$, $q = \operatorname{card}(J)$, $r = \operatorname{card}(K)$, and

$$Z = \left\{ (z_i)_{i\in I\cup J\cup K} \in \mathbb{R}^{p+q+r} \;\middle|\; \exists x \in C, \begin{array}{l} \forall i \in I, \ f_i(x) < z_i \\ \forall i \in J \cup K, \ g_i(x) = z_i \end{array} \right\}$$

The set Z is convex, nonempty and $Z \cap (\mathbb{R}_-^{p+q} \times \{0_{\mathbb{R}^r}\}) = \emptyset$. From Proposition 3.2.6 one can properly separate $(\mathbb{R}_-^{p+q} \times \{0_{\mathbb{R}^r}\})$ from Z by a hyperplane which does not contain Z, i.e. there exists real scalars $(\lambda_i)_{i\in I}$, $(\mu_i)_{i\in J\cup K}$, all of them not being equal to zero, which verify:

$$\sum_{i\in I} \lambda_i z_i + \sum_{i\in J\cup K} \mu_i z_i \geq \sum_{i\in I} \lambda_i \zeta_i + \sum_{i\in J} \mu_i \zeta_i, \ \forall z \in Z, \ \forall \zeta \in \mathbb{R}_-^{p+q}$$

and there exists $z \in Z$ such that $\sum_{i\in I}\lambda_i z_i + \sum_{i\in J\cup K}\mu_i z_i > 0$.
If for some i, $\lambda_i < 0$, then letting z_i tend to $+\infty$, we get a contradiction. Thus $\lambda_i \geq 0$, $\forall i \in I$. If for some $i \in J$, one has $\mu_i < 0$, then letting ζ_i tend to $-\infty$, we have another contradiction. Hence, $\mu_i \geq 0$, $\forall i \in J$. Let $\varepsilon > 0$ and $x \in C$. Define $z \in Z$ by

$$\begin{cases} z_i = f_i(x) + \varepsilon, \ \forall i \in I, \\ z_i = g_i(x), \qquad \forall i \in J \cup K. \end{cases}$$

We have $\sum_{i\in I}\lambda_i f_i(x) + \sum_{i\in J\cup K}\mu_i g_i(x) + \varepsilon\sum_{i\in I}\lambda_i \geq 0$. Let ε tend to zero. We obtain that $\sum_{i\in I}\lambda_i f_i(x) + \sum_{i\in J\cup K}\mu_i g_i(x) \geq 0$, $\forall x \in C$.
To end the proof, it remains to show that at least one of the $(\lambda_i)_{i\in I}$ is strictly positive. Assume the contrary. We then have $\sum_{i\in J\cup K}\mu_i g_i(x) \geq 0$, $\forall x \in C$ and hence $\sum_{i\in J\cup K}\mu_i g_i(x_0) = 0$. Since x_0 is in ri(C), the affine function $\sum_{i\in J\cup K}\mu_i g_i$ has a minimum in the relative interior of C. One can easily check that it must be equal to zero on C. But since $\sum_{i\in I}\lambda_i z_i + \sum_{i\in J\cup K}\mu_i z_i > 0$ for some $z \in Z$, and since all the $(\lambda_i)_{i\in I}$ are equal to zero, there exists $x \in C$ such that $\sum_{i\in J\cup K}\mu_i g_i(x) > 0$. That contradicts that $\sum_{i\in J\cup K}\mu_i g_i$ is equal to zero on C. □

Let us now introduce two definitions.
Let C be a nonempty convex set of $\mathbb{R}^n$. The *normal cone* at $x \in C$ is the cone :

$$N_C(x) = \{v \in \mathbb{R}^n \mid v\cdot(z-x) \leq 0, \ \forall z \in C\}.$$

Note that with the previous notations, $N_C(x) = (K(C - \{x\}))^\circ$.

The *indicator* of C, δ_C, is the convex function which is equal to 0 on C and to $+\infty$ outside of C.
Figure 7.1 will highlight the notion of normal cone. When the set C is defined as intersection of a finite number of level sets, simple characterizations of the normal cone at a point $x \in C$ will be given by Corollaries 7.2.2 and 7.2.3.

Before stating Proposition 7.2.1, we need to state a lemma which relates $N_C(x)$ and $\partial\delta_C(x)$, and to define Slater Condition.

Lemma 7.2.2. *Let C be a nonempty convex set of $\mathbb{R}^n$. Then*

$$N_C(x) = \partial\delta_C(x), \ \forall x \in \mathbb{R}^n.$$

Proof. Left for the reader (see Exercise 7.1). □

The Problem (P) satisfies *Slater Condition (S)* if there exists $x_0 \in$ ri(dom(P)) such that:

$$(S) \qquad \begin{cases} f_i(x_0) < 0, \ \forall i \in I \\ g_i(x_0) \leq 0, \ \forall i \in J \\ g_i(x_0) = 0, \ \forall i \in K. \end{cases}$$

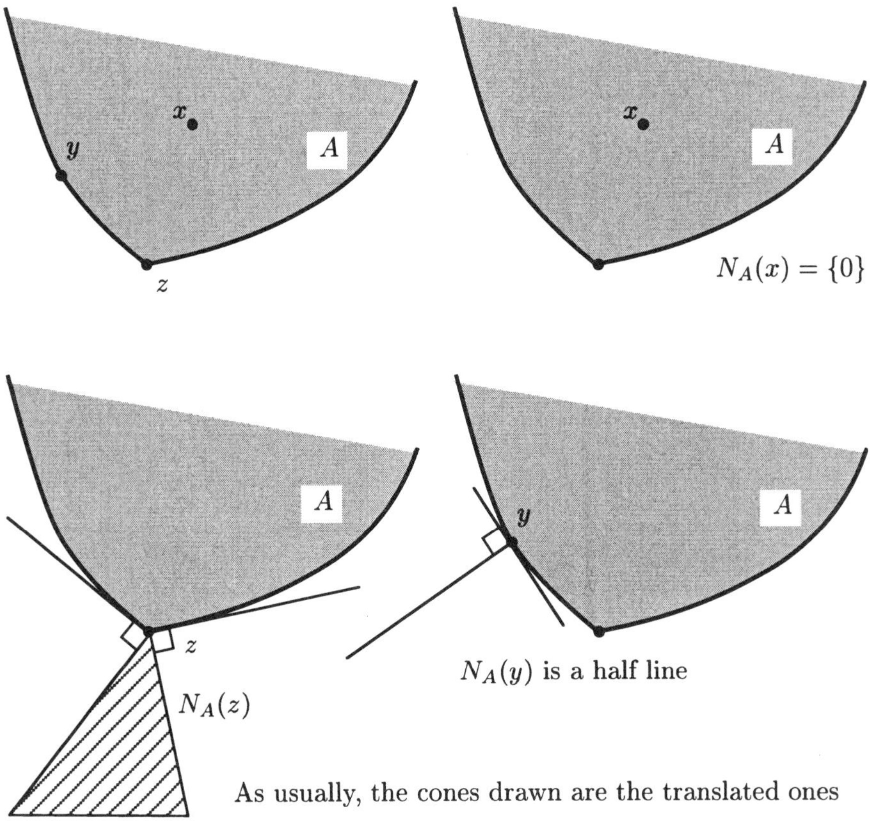

Fig. 7.1. Normal cone

Proposition 7.2.1. *Consider Problem (P). Assume that (P) satisfies Slater Condition (S). If (P) has an optimal solution* $\overline{x}$*, then there exists scalars* $(\lambda_i)_{i\in I}$, $(\mu_i)_{i\in J\cup K}$ *such that:*
(i) $\forall i \in I,\ \lambda_i \geq 0, \lambda_i f_i(\overline{x}) = 0,$
$\forall i \in J,\ \mu_i \geq 0, \mu_i g_i(\overline{x}) = 0.$
(ii) $0 \in \partial f_0(\overline{x}) + \sum_{i\in I} \lambda_i \partial f_i(\overline{x}) + \sum_{i\in J\cup K} \mu_i\{\nabla g_i(\overline{x})\} + N_C(\overline{x}),$
where $C = \mathrm{dom}(P)$ *and* $N_C(\overline{x})$ *is the normal cone at* $\overline{x}$ *to* C.

Proof. Let $\alpha = f_0(\overline{x})$. Then the following system has no solution in C.

$$\begin{cases} f_0(x) < \alpha, \\ f_i(x) < 0, \ \forall i \in I, \\ g_i(x) \leq 0, \ \forall i \in J, \\ g_i(x) = 0, \ \forall i \in K. \end{cases}$$

The Slater condition allows us to apply Lemma 7.2.1.
There exist $\lambda_0, (\lambda_i)_{i\in I}, (\mu_i)_{i\in J\cup K}$ such that $\lambda_0 \geq 0, \lambda_i \geq 0,\ \forall i \in I, \mu_i \geq 0,$

$\forall i \in J$, at least one of the $\lambda_0, (\lambda_i)_{i\in I}$ is strictly positive, and

$$\lambda_0(f_0(x) - \alpha) + \sum_{i\in I} \lambda_i f_i(x) + \sum_{i\in J\cup K} \mu_i g_i(x) \geq 0, \ \forall x \in C. \tag{7.1}$$

We claim that $\lambda_0 > 0$. Indeed, if $\lambda_0 = 0$, then $\forall x \in C$, $\sum_{i\in I} \lambda_i f_i(x) + \sum_{i\in J\cup K} \mu_i g_i(x) \geq 0$. Moreover, since there exists at least one $i \in I$ such that $\lambda_i > 0$, we deduce $\sum_{i\in I} \lambda_i f_i(x_0) + \sum_{i\in J\cup K} \mu_i g_i(x_0) < 0$, a contradiction. Thus $\lambda_0 > 0$ and one can suppose $\lambda_0 = 1$.
Define the convex function h by

$$h(x) = f_0(x) + \sum_{i\in I}(\lambda_i f_i)(x) + \sum_{i\in J\cup K} \mu_i g_i(x), \ \forall x \in \mathbb{R}^n.$$

The inequality (7.1) can be equivalently rewritten as $h(x) \geq \alpha$, $\forall x \in C$. But $h(\overline{x}) - \alpha = \sum_{i\in I} \lambda_i f_i(\overline{x}) + \sum_{i\in J\cup K} \mu_i g_i(\overline{x}) \leq 0$. Hence, $\alpha = h(\overline{x})$. We can also write: $h(x) + \delta_C(x) \geq h(\overline{x}) + \delta_C(\overline{x})$, $\forall x \in \mathbb{R}^n$. By Remark 7.2.1, $0 \in \partial(h + \delta_C)(\overline{x})$. Since (Exercise 1.26)

$$\emptyset \neq \mathrm{ri}(C) = \mathrm{ri}(\mathrm{dom}(f_0)) \cap (\cap_{i\in I} \mathrm{ri}(\mathrm{dom}(f_i))) \subset \mathrm{ri}(\mathrm{dom}(h)),$$

and since $C = \mathrm{dom}(\delta_C), \mathrm{dom}(g_i) = \mathbb{R}^n$, one can apply Lemma 7.2.2 and Proposition 6.5.5 to successively write:

$$\partial(h + \delta_C)(\overline{x}) = \partial h(\overline{x}) + \partial \delta_C(\overline{x}) = \partial h(\overline{x}) + N_C(\overline{x}),$$

and

$$\partial h(\overline{x}) = \partial f_0(\overline{x}) + \sum_{i\in I} \lambda_i \partial f_i(\overline{x}) + \sum_{i\in J\cup K} \mu_i\{\nabla g_i(\overline{x})\}$$

since $\partial g_i(\overline{x}) = \{\nabla g_i(\overline{x})\}$ and since by our rule, if $\lambda_i = 0$, then $\lambda_i \partial f_i(\overline{x}) = \{0\}$, even if $\partial f_i(\overline{x}) = \emptyset$. We have proved Assertion (ii).
Since $\sum_{i\in I} \lambda_i f_i(\overline{x}) + \sum_{i\in J\cup K} \mu_i g_i(\overline{x}) = 0$ and since $\overline{x}$ is feasible, one has $\forall i \in I$, $\lambda_i f_i(\overline{x}) = 0$ and $\forall i \in J$, $\mu_i g_i(\overline{x}) = 0$. We have proved Assertion (i). The proof is now complete. □

The real numbers $(\lambda_i)_{i\in I}, (\mu_i)_{i\in J\cup K}$ are called *Lagrange parameters*, *Lagrange multipliers* or *Kuhn-Tucker coefficients* or more simply *multipliers* of Problem (P).
The following conditions (i), (ii) and (iii) are called *Kuhn-Tucker Conditions* for Problem (P).
We say that $\overline{x}$, $(\lambda_i)_{i\in I}$, $(\mu_i)_{i\in J\cup K}$ satisfy Kuhn-Tucker Conditions of Problem (P) if they satisfy Conditions (i), (ii) and (iii). Observe that in this case, $\overline{x} \in \mathrm{dom}(f_0)$.

(i) $\forall i \in I$, $\lambda_i \geq 0, f_i(\overline{x}) \leq 0, \lambda_i f_i(\overline{x}) = 0$,
$\forall i \in J$, $\mu_i \geq 0, g_i(\overline{x}) \leq 0, \mu_i g_i(\overline{x}) = 0$.
(ii) $\forall i \in K$, $g_i(\overline{x}) = 0$.

(iii) $0 \in \partial f_0(\overline{x}) + \sum_{i\in I} \lambda_i \partial f_i(\overline{x}) + \sum_{i\in J\cup K} \mu_i\{\nabla g_i(\overline{x})\} + N_C(\overline{x})$, where $C = \text{dom}(P)$.

As particular case of Proposition 7.2.1, we get the following result when the problem is without convex constraints.

Corollary 7.2.1. *Consider Problem (P) without convex constraints, i.e.*

$$\min f_0(x) \text{ under the constraints } \begin{cases} g_i(x) \leq 0, & \forall i \in J \\ g_i(x) = 0, & \forall i \in K. \end{cases}$$

where $f_0 : \mathbb{R}^n \to \mathbb{R} \cup \{+\infty\}$ *is convex, and for all* $i \in J \cup K$, g_i *is affine.*
For this problem, Slater Condition is: $\exists x_0 \in \text{ri}(\text{dom}(f_0))$ *such that* $g_i(x_0) \leq 0$, $\forall i \in J$, *and* $g_i(x_0) = 0$, $\forall i \in K$.
If (P) has an optimal solution $\overline{x}$, *then there exist scalars* $(\mu_i)_{i\in J\cup K}$ *verifying:*
(i) $\forall i \in J$, $\mu_i \geq 0, \mu_i g_i(\overline{x}) = 0$,
(ii) $0 \in \partial f_0(\overline{x}) + \sum_{i\in J\cup K} \mu_i\{\nabla g_i(\overline{x})\} + N_C(\overline{x})$, *where* $C = \text{dom}(f_0)$.

Proof. Obvious □

Remark 7.2.2. (i) In Proposition 7.2.1 , if the functions f_i, for $i \in I$ and f are real-valued, then $C = \mathbb{R}^n$ and $N_C(\overline{x}) = \{0\}$.
(ii) If in Proposition 7.2.1, the functions f_i, for $i \in I$ and f are differentiable then Condition (ii) becomes $\nabla f_0(\overline{x}) + \sum_{i\in I} \lambda_i \nabla f_i(\overline{x}) + \sum_{i\in J\cup K} \mu_i \nabla g_i(\overline{x}) \in -N_C(\overline{x})$. If the functions f_i, for $i \in I$ and f are real-valued then we obtain the well-known result

$$\nabla f_0(\overline{x}) + \sum_{i\in I} \lambda_i \nabla f_i(\overline{x}) + \sum_{i\in J\cup K} \mu_i \nabla g_i(\overline{x}) = 0.$$

(iii) The Slater condition is *very important* even in the one-dimensional case. Consider the following example:

$$\min\{f(x) = x \mid x^2 \leq 0\}.$$

The problem has a unique solution $\overline{x} = 0$. The Slater condition is not satisfied. There exists no $\lambda \geq 0$ such that $0 = \nabla f(\overline{x}) + \lambda(2\overline{x}) = \nabla f(\overline{x})$ because $\nabla f(\overline{x}) = 1$.
(iv) It is important to notice that Slater Condition is not necessary to obtain Kuhn-Tucker Conditions. In the previous example, Slater Condition is not satisfied and there is no Kuhn-Tucker coefficient. Now replace this problem by an *equivalent* problem which is $\min\{f(x) = x \mid g(x) = |x| \leq 0\}$. As before, Slater Condition does not hold. The unique solution is always $\overline{x} = 0$. Observe that $\partial g(0) = [-1, +1]$ (see Proposition 6.7.4 or Exercise 6.8). Let $\lambda = 1$, we have successively $g(0) = 0, \lambda g(0) = 0$ and $0 \in \{\nabla f(0)\} + \lambda \partial g(0)$, where $\nabla f(0) = 1$. In other words, Kuhn-Tucker Conditions hold.
(vii) In the previous example, one can check that Kuhn-Tucker Conditions are sufficient for 0 to be a solution. This result is quite general as it will be proved in the next proposition.

Proposition 7.2.2. *Let* $\overline{x}, (\lambda_i)_{i\in I}, (\mu_i)_{i\in J\cup K}$ *verify Kuhn-Tucker Conditions for Problem* (P). *Then* $\overline{x}$ *is a solution to* (P).

Proof. Let $\forall x \in \mathbb{R}^n$, $h(x) = f_0(x) + \sum_{i\in I}(\lambda_i f_i)(x) + \sum_{i\in J\cup K}\mu_i g_i(x)$ (with the rule that $(\lambda_i f_i) = 0$ if $\lambda_i = 0$). Since $\mathrm{ri}(\mathrm{dom}(P)) \neq \emptyset$, we can apply Proposition 6.5.5 to write

$$\forall x \in \mathbb{R}^n,\ \partial h(x) = \partial f_0(x) + \sum_{i\in I}\partial(\lambda_i f_i)(x) + \sum_{i\in J\cup K}\mu_i\{\nabla g_i(x)\}.$$

We have the following condition (Kuhn-Tucker Condition (iii))

$$\begin{aligned} 0 &\in \partial f_0(\overline{x}) + \sum_{i\in I}\lambda_i\partial f_i(\overline{x}) + \sum_{i\in J\cup K}\mu_i\{\nabla g_i(\overline{x})\} + N_C(\overline{x}) \\ &= \partial f_0(\overline{x}) + \sum_{i\in I}\partial(\lambda_i f_i)(\overline{x}) + \sum_{i\in J\cup K}\mu_i\{\nabla g_i(\overline{x})\} + N_C(\overline{x}). \end{aligned}$$

Since $N_C(\overline{x}) = \partial\delta_C(\overline{x})$ and since $\delta_C(\overline{x}) = 0$, the following inequality holds for every $x \in \mathbb{R}^n$:

$$\begin{aligned} f_0(\overline{x}) + \sum_{i\in I}(\lambda_i f_i)(\overline{x}) + \sum_{i\in J\cup K}\mu_i g_i(\overline{x}) &\leq \\ f_0(x) + \sum_{i\in I}(\lambda_i f_i)(x) + \sum_{i\in J\cup K}\mu_i g_i(x) + \delta_C(x). \end{aligned}$$

We will prove that $f_0(x) \geq f_0(\overline{x})$, for all feasible point x. Indeed, if $x \notin \mathrm{dom}(f_0)$, then $f_0(x) = +\infty > f_0(\overline{x})$. If $x \in \mathrm{dom}(f_0)$ and if it verifies the constraints of Problem (P), then $x \in C$ and $\delta_C(x) = 0$. The Kuhn-Tucker conditions (i) and (ii) and the constraints that x must satisfy, imply that $f_0(x) \geq f_0(\overline{x})$. □

Theorem 7.2.1 (Kuhn-Tucker). *Assume that Slater Condition is satisfied for Problem* (P). *Then* $\overline{x}$ *is a solution to* (P) *if, and only if, there exists coefficients* $(\lambda_i)_{i\in I}, (\mu_i)_{i\in J\cup K}$ *which, together with* $\overline{x}$, *satisfy Kuhn-Tucker Conditions for Problem* (P).

Proof. The statement follows from Proposition 7.2.1 and Proposition 7.2.2. □

An interesting case is when the functions f_i, for $i \in I$ are real-valued on the domain of f_0 (see Rockafellar [51]). We will show that in this case, we can drop the term $N_C(\overline{x})$ in Kuhn-Tucker Condition (iii).

Proposition 7.2.3. *Consider Problem* (P). *Assume the functions* $(f_i)_{i\in I}$ *are real-valued on* $C = \mathrm{dom}(f_0)$. *Assume that* (P) *satisfies Condition* (S')

$$\exists x_0 \in \mathrm{ri}(C) \textit{ such that } \begin{cases} f_i(x_0) < 0, & \forall i \in I \\ g_i(x_0) \leq 0, & \forall i \in J \\ g_i(x_0) = 0, & \forall i \in K. \end{cases}$$

If (P) has a solution $\overline{x}$, then there exists scalars $(\lambda_i)_{i\in I}, (\mu_i)_{i\in J\cup K}$ such that:
(i) $\forall i \in I$, $\lambda_i \geq 0, \lambda_i f_i(\overline{x}) = 0$,
$\forall i \in J$, $\mu_i \geq 0, \mu_i g_i(\overline{x}) = 0$.
(ii) $0 \in \partial f_0(\overline{x}) + \sum_{i\in I} \lambda_i \partial f_i(\overline{x}) + \sum_{i\in J\cup K} \mu_i\{\nabla g_i(\overline{x})\}$.

Proof. Since, $C = \mathrm{dom}(P)$. Condition (S') is Slater Condition for Problem (P). We can apply Proposition 7.2.1. There exists scalars $(\lambda_i)_{i\in I}, (\mu_i)_{i\in J\cup K}$ such that:
(i) $\forall i \in I$, $\lambda_i \geq 0, \lambda_i f_i(\overline{x}) = 0$,
$\forall i \in J$, $\mu_i \geq 0, \mu_i g_i(\overline{x}) = 0$.
(ii) $0 \in \partial f_0(\overline{x}) + \sum_{i\in I} \lambda_i \partial f_i(\overline{x}) + \sum_{i\in J\cup K} \mu_i\{\nabla g_i(\overline{x})\} + N_C(\overline{x})$,
where $N_C(\overline{x})$ is the normal cone at $\overline{x}$ to C. Rewrite Condition (ii) as:

$$\forall x \in \mathbb{R}^n,\ f_0(x) + \sum_{i\in I}(\lambda_i f_i)(x) + \sum_{i\in J\cup K} \mu_i g_i(x) + \delta_C(x) \geq$$
$$f_0(\overline{x}) + \sum_{i\in I}(\lambda_i f_i)(\overline{x}) + \sum_{i\in J\cup K} \mu_i g_i(\overline{x}) + \delta_C(\overline{x}).$$

If $x \in C$ then

$$f_0(x) + \sum_{i\in I}(\lambda_i f_i)(x) + \sum_{i\in J\cup K} \mu_i g_i(x) \geq f_0(\overline{x}) + \sum_{i\in I}(\lambda_i f_i)(\overline{x}) + \sum_{i\in J\cup K} \mu_i g_i(\overline{x}).$$

If $x \notin C$ then $f_0(x) = +\infty$ and the previous inequality still holds. Thus, actually this inequality is true for every $x \in \mathbb{R}^n$. Define the function $h : \mathbb{R}^n \to \mathbb{R} \cup \{\infty\}$ by

$$\forall x \in \mathbb{R}^n,\ h(x) = f_0(x) + \sum_{i\in I}(\lambda_i f_i)(x) + \sum_{i\in J\cup K} \mu_i g_i(x).$$

We have $\forall x \in \mathbb{R}^n$, $h(x) \geq h(\overline{x})$ or equivalently $0 \in \partial h(\overline{x})$. Since $\mathrm{ri}(C) \neq \emptyset$, from Proposition 6.5.5,

$$\partial h(\overline{x}) = \partial f_0(\overline{x}) + \sum_{i\in I} \lambda_i \partial f_i(\overline{x}) + \sum_{i\in J\cup K} \mu_i\{\nabla g_i(\overline{x})\}.$$ □

Corollary 7.2.2 (Normal cones). *(i) Let f be a convex function from $\mathbb{R}^n$ into $\mathbb{R}$. Let $C = \{x \mid f(x) \leq 0\}$. Let $\overline{x} \in C$. We have:*
(i) If $f(\overline{x}) < 0$ then $N_C(\overline{x}) = \{0\}$.
(ii) If $f(\overline{x}) = 0$ and if $0 \notin \partial f(\overline{x})$ then $N_C(\overline{x}) = \{\lambda \partial f(\overline{x}) \mid \lambda \geq 0\}$.
Hence, if f is non-constant affine , i.e. $f(x) = a \cdot x + b$ with $a \neq 0$, then

$$N_C(\overline{x}) = \begin{cases} \{0\} & \text{if } a\cdot\overline{x} + b < 0 \\ \{\lambda a \mid \lambda \geq 0\} & \text{if } a\cdot\overline{x} + b = 0. \end{cases}$$

We also have: $N_H(\overline{x}) = \{\lambda a \mid \lambda \in \mathbb{R}\}$, where $H = \{x \mid a\cdot x + b = 0\}$.

Proof. Observe that $v \in N_C(\overline{x})$ if, and only if, $\overline{x}$ solves the following problem $(P) : \min\{-v \cdot x \mid f(x) \leq 0\}$.
(i) Assume $f(\overline{x}) < 0$ and $v \in N_C(\overline{x})$. Hence Slater Condition holds for Problem (P). There exists $\lambda \geq 0$ such that $\lambda f(\overline{x}) = 0$ and $0 \in \{-v\} + \lambda\partial f(\overline{x})$. Since $f(\overline{x}) < 0$, one has $\lambda = 0$ and the conclusion holds.

(ii) Assume now $f(\overline{x}) = 0$ and $v \in N_C(\overline{x})$. We claim that there exists x_0 with $f(x_0) < 0$. If not, $\forall x \in \mathbb{R}^n$, $f(x) \geq f(\overline{x})$ in contradiction with the assumption that $0 \notin \partial f(\overline{x})$. Thus Slater Condition holds for Problem (P). From Proposition 7.2.3, $0 \in \{-v\} + \lambda\partial f(\overline{x}), \lambda \geq 0$, or, equivalently, $v \in \{\lambda\partial f(\overline{x}) \mid \lambda \geq 0\}$.
Conversely, if $v \in \{\lambda\partial f(\overline{x})\}$ with $\lambda \geq 0$, or equivalently $0 \in \{-v\} + \lambda\partial f(\overline{x})$, since $f(\overline{x}) = 0$, we have Kuhn-Tucker Conditions for Problem (P). From Proposition 7.2.2, $\overline{x}$ solves (P). Hence, $v \in N_C(\overline{x})$.
The application of these results to the case where f is affine is routine. □

Corollary 7.2.3 (Normal cones of intersection). *Let f_i, for $i \in I$ be a convex, non-affine function from $\mathbb{R}^n$ into $\mathbb{R}$, g_i, for $i \in J \cup K$ be an affine function. Let $C_i = \{x \mid f_i(x) \leq 0\}$ and $C_i' = \{x \mid f_i(x) < 0\}$, for $i \in I$, $D_i = \{x \mid g_i(x) \leq 0\}$, for $i \in J$, $E_i = \{x \mid g_i(x) = 0\}$, for $i \in K$, and $C = (\cap_{i\in I} C_i) \cap (\cap_{i\in J} D_i) \cap (\cap_{i\in K} E_i)$. Assume that:*

$$\left(\bigcap\nolimits_{i\in I} C_i'\right) \cap \left(\bigcap\nolimits_{i\in J} D_i\right) \cap \left(\bigcap\nolimits_{i\in K} E_i\right) \neq \emptyset,$$

then for all $\overline{x} \in C$, one has:

$$N_C(\overline{x}) = \sum_{i\in I} N_{C_i}(\overline{x}) + \sum_{i\in J} N_{D_i}(\overline{x}) + \sum_{i\in K} N_{E_i}(\overline{x}).$$

Proof. It is obvious that $v \in N_C(\overline{x})$ if, and only if, $\overline{x}$ solves Problem (P): $\min\{-v \cdot x \mid x \in C\}$. To assume $(\cap_{i\in I} C_i\prime) \cap (\cap_{i\in J} D_i) \cap (\cap_{i\in K} E_i) \neq \emptyset$ is equivalent to assume that Slater Condition holds for Problem (P).
Let $v \in N_C(\overline{x})$. We apply Proposition 7.2.3. There exist nonnegative numbers $(\lambda_i)_{i\in I}$, $(\mu_i)_{i\in J}$, and real numbers $(\mu_i)_{i\in K}$ such that:
(i) $\lambda_i f_i(\overline{x}) = 0, \ \forall i \in I$,
$\mu_i g_i(\overline{x}) = 0, \ \forall i \in J$,
(ii) $0 \in \{-v\} + \sum_{i\in I} \lambda_i \partial f_i(\overline{x}) + \sum_{i\in J\cup K} \mu_i \nabla g_i(\overline{x})$.

Let $I(\overline{x}) = \{i \in I \mid f_i(\overline{x}) = 0\}$ and $J(\overline{x}) = \{i \in J \mid g_i(\overline{x}) = 0\}$. If $i \notin I(\overline{x})$, we deduce from the previous corollary that $N_{C_i}(\overline{x}) = \{0\}$, and by (i) $\lambda_i = 0$. Similarly, if $i \notin J(\overline{x})$ we have $N_{D_i}(\overline{x}) = \{0\}$ and $\mu_i = 0$.
Observe that the condition $(\cap_{i\in I}(C_i')) \cap (\cap_{i\in J}^q D_i) \cap (\cap_{i\in K} E_i) \neq \emptyset$ implies that for $i \in I(\overline{x})$, $\overline{x}$ is not a minimizer of f_i and thus $0 \notin \partial f_i(\overline{x})$. Hence (ii) is equivalent to $v \in \sum_{i\in I} N_{C_i}(\overline{x}) + \sum_{i\in J} N_{D_i}(\overline{x}) + \sum_{i\in K} N_{E_i}(\overline{x})$. We have proved that $N_C(\overline{x}) \subset \sum_{i\in I} N_{C_i}(\overline{x}) + \sum_{i\in J} N_{D_i}(\overline{x}) + \sum_{i\in K} N_{E_i}(\overline{x})$.

Let us prove the converse. Let

$$v \in \sum_{i\in I} N_{C_i}(\overline{x}) + \sum_{i\in J} N_{D_i}(\overline{x}) + \sum_{i\in K} N_{E_i}(\overline{x}).$$

Consider the sets $I(\overline{x}), J(\overline{x})$ defined above. Using the previous corollary, we get the existence of scalars $(\lambda_i)_{i\in I(\overline{x})}, (\mu_i)_{i\in J(\overline{x})\cup K}$ such that

$$v \in \sum_{i\in I(\overline{x})} \lambda_i \partial f_i(\overline{x}) + \sum_{i\in J(\overline{x})} \mu_i\{\nabla g_i(\overline{x})\} + \sum_{i\in K} \mu_i\{\nabla g_i(\overline{x})\},$$

and satisfying $\forall i \in I(\overline{x})$, $\lambda_i \geq 0$, $\forall i \in J(\overline{x})$, $\mu_i \geq 0$. Define $\lambda_i = 0$, $\forall i \in I \setminus I(\overline{x})$, $\mu_i = 0$, $\forall i \in J \setminus J(\overline{x})$. One can easily check that Kuhn-Tucker Conditions for Problem (P) hold at $\overline{x}$ with $(\lambda_i)_{i\in I}, (\mu_i)_{i\in J\cup K}$ as Kuhn-Tucker Coefficients. By Proposition 7.2.2, $\overline{x}$ is a solution to Problem (P). Thus, v belongs to $N_C(\overline{x})$. The proof is complete. □

Remark 7.2.3. (i) The identity between the normal cone $N_C(\overline{x})$ and the cone generated by $\partial f(\overline{x})$, where $C = \{x \mid f(x) \leq 0\}$ requires that $0 \notin \partial f(\overline{x})$. We give here an example where $N_C(\overline{x})$ is different from the cone generated by $\partial f(\overline{x})$.
Consider the convex function in $\mathbb{R}$: $f(x) = x^2$. Then $C = \{0\}$. We have $N_C(0) = \mathbb{R}$ while $\partial f(0) = \{0\}$. The cone it generates is $\{0\}$.
(ii) In the previous corollary, condition $(\cap_{i\in I} C_i') \cap (\cap_{i\in J} D_i) \cap (\cap_{i\in K} E_i) \neq \emptyset$ is also very important. We give an example when it does not hold and the statement is false. Let

$$C_1 = \{(x,y) \in \mathbb{R}^2 \mid (x-1)^2 + y^2 \leq 1\}$$

and

$$D = \{(x,y) \in \mathbb{R}^2 \mid x \leq 0\}.$$

We have $C_1' \cap D = \emptyset$, $C = C_1 \cap D = \{(0,0)\}$. Hence $N_C((0,0)) = \mathbb{R}^2$. But $N_{C_1}((0,0)) + N_D((0,0)) = \{\lambda(1,0) \mid \lambda \in \mathbb{R}\}$.

We now show how the Kuhn-Tucker coefficients and the optimal solutions can be characterize in terms of the saddle-point of a certain concave-convex function.

7.2.2 Lagrangian of Problem (P)

The *Lagrangian* of Problem (P) is the function $L : \mathbb{R}^p_+ \times \mathbb{R}^q_+ \times \mathbb{R}^{r-q} \times \mathbb{R}^n \to \mathbb{R} \cup \{+\infty\}$ defined by for all $(\lambda, \mu, x) = ((\lambda_i)_{i\in I}, (\mu_i)_{i\in J}, (\mu_i)_{i\in K}, x)$,

$$L(\lambda,\mu,x) = f_0(x) + \sum_{i\in I}(\lambda_i f_i)(x) + \sum_{i\in J} \mu_i g_i(x) + \sum_{i\in K} \mu_i g_i(x).$$

where $p = \text{card}(I), q = \text{card}(J)$ and $r = \text{card}(J \cup K)$.
We say that $(\overline{\lambda}, \overline{\mu}, \overline{x})$ is a *saddle-point* of L if for all $\forall(\lambda,\mu) \in \mathbb{R}^p_+ \times \mathbb{R}^q_+ \times \mathbb{R}^{r-q}, \forall x \in \mathbb{R}^n$,

$$L(\lambda,\mu,\overline{x}) \leq L(\overline{\lambda},\overline{\mu},\overline{x}) \leq L(\overline{\lambda},\overline{\mu},x).$$

Observe that since $\text{dom}(P) \neq \emptyset$, if $(\overline{\lambda}, \overline{\mu}, \overline{x})$ is a *saddle-point* of L, then $L(\overline{\lambda},\overline{\mu},\overline{x}) \in \mathbb{R}$. Indeed, we have $f_0(\overline{x}) = L(0,0,\overline{x}) \leq L(\overline{\lambda},\overline{\mu},\overline{x}) \leq L(\overline{\lambda},\overline{\mu},x) < +\infty$, for any $x \in \text{dom}(P)$.

Theorem 7.2.2. *(i) If $(\overline{\lambda}, \overline{\mu}, \overline{x})$ is a saddle-point of L, then $\overline{x}$ is a solution to (P) and $(\overline{\lambda}, \overline{\mu})$ are the Kuhn-Tucker coefficients associated with $\overline{x}$. Moreover, $L(\overline{\lambda}, \overline{\mu}, \overline{x}) = f_0(\overline{x})$.*
(ii) Assume moreover $\mathrm{dom}(f_0) = \mathrm{dom}(P)$. If $(\overline{\lambda}, \overline{\mu}, \overline{x})$ satisfies Kuhn-Tucker Conditions for Problem (P), then $(\overline{\lambda}, \overline{\mu}, \overline{x})$ is a saddle-point of L.

Proof. (i) Let $(\overline{\lambda}, \overline{\mu}, \overline{x})$ be a saddle-point of L and $C = \mathrm{dom}(P)$. Since $\mathrm{ri}(C) \neq \emptyset$, we have $L(\overline{\lambda}, \overline{\mu}, \overline{x}) \in \mathbb{R}$. By definition of a saddle-point, one has:

$$L(\overline{\lambda}, \overline{\mu}, \overline{x}) = \sup_{(\lambda,\mu)\in\mathbb{R}^p_+\times\mathbb{R}^q_+\times\mathbb{R}^{r-q}} L(\lambda, \mu, \overline{x})$$

$$= \sup_{(\lambda,\mu)\in\mathbb{R}^p_+\times\mathbb{R}^q_+\times\mathbb{R}^{r-q}} f_0(\overline{x}) + \sum_{i\in I} \lambda_i f_i(\overline{x}) + \sum_{i\in J\cup K} \mu_i g_i(\overline{x}).$$

If for some i, we have $f_i(\overline{x}) > 0$, the second member of the inequality equals $+\infty$. Hence $f_i(\overline{x}) \leq 0$, $\forall i$. If $f_i(\overline{x}) < 0$, the supremum is reached by $\overline{\lambda}_i = 0$. The same assertion holds for $g_i(\overline{x}), i \in J$. For $i \in K$, if $g_i(\overline{x}) \neq 0$ the second member equals $+\infty$. Hence $g_i(\overline{x}) = 0, \forall i \in K$.
One the other hand, we have:

$$L(\overline{\lambda}, \overline{\mu}, \overline{x}) = \inf_{x\in\mathbb{R}^n} L(\overline{\lambda}, \overline{\mu}, x) = \inf_{x\in C} L(\overline{\lambda}, \overline{\mu}, x) = \inf_{x\in\mathbb{R}^n} (L(\overline{\lambda}, \overline{\mu}, x) + \delta_C(x)).$$

Since L is convex with respect to x, one has: $0 \in \partial_x(L(\overline{\lambda}, \overline{\mu}, .) + \delta_C(.))(\overline{x})$. The set C is included in the domain of the function : $\mathbb{R}^n \to \mathbb{R} \cup \{+\infty\}$ defined by $x \to L(\overline{\lambda}, \overline{\mu}, x)$. Since $\mathrm{ri}(\mathrm{dom}(P)) \neq \emptyset$, the relative interior of the domain of the function $x \to L(\overline{\lambda}, \overline{\mu}, x)$ is also nonempty. One can apply Proposition 6.5.5 to obtain

$$0 \in \partial f_0(\overline{x}) + \sum_{i\in I} \partial(\overline{\lambda}_i f_i)(\overline{x}) + \sum_{i\in J\cup K} \overline{\mu}_i\{\nabla g_(\overline{x})\} + N_C(\overline{x})$$
$$= \partial f_0(\overline{x}) + \sum_{i\in I} \overline{\lambda}_i \partial f_i(\overline{x}) + \sum_{i\in J\cup K} \overline{\mu}_i\{\nabla g_(\overline{x})\} + N_C(\overline{x}),$$

with our rules on (λf) and $\lambda\partial f(x)$. We obtain Kuhn-Tucker Conditions. By Proposition 7.2.2, $\overline{x}$ is a solution to Problem (P). One can easily check that $L(\overline{\lambda}, \overline{\mu}, \overline{x}) = f_0(\overline{x})$.

(ii) Conversely, assume moreover that $C = \mathrm{dom}(f_0)$ and that $(\overline{\lambda}, \overline{\mu}, \overline{x})$ satisfy Kuhn-Tucker Conditions for Problem (P). In particular we have :

$$\forall x \in \mathbb{R}^n,\ f_0(x) + \sum_{i\in I}(\overline{\lambda}_i f_i)(x) + \sum_{i\in J\cup K} \overline{\mu}_i g_i(x) + \delta_C(x) \geq$$
$$f_0(\overline{x}) + \sum_{i\in I}(\overline{\lambda}_i f_i)(\overline{x}) + \sum_{i\in J\cup K} \overline{\mu}_i g_i(\overline{x}).$$

If $x \in C$ then $\delta_C(x) = 0$ and

$$f_0(x) + \sum_{i\in I}(\overline{\lambda}_i f_i)(x) + \sum_{i\in J\cup K} \overline{\mu}_i g_i(x) \geq f_0(\overline{x}) + \sum_{i\in I}(\overline{\lambda}_i f_i)(\overline{x}) + \sum_{i\in J\cup K} \overline{\mu}_i g_i(\overline{x}).$$

If $x \notin C$ then $f_0(x) = +\infty$ and the previous inequality also holds. In other words, $L(\overline{\lambda}, \overline{\mu}, x) \geq L(\overline{\lambda}, \overline{\mu}, \overline{x})$, $\forall x \in \mathbb{R}^n$.
But $f_i(\overline{x}) \leq 0$, $\forall i \in I$ and $g_i(\overline{x}) \leq 0$, $\forall i \in J \cup K$. Thus $\sum_{i\in I} \lambda_i f_i(\overline{x}) + \sum_{i\in J\cup K} \mu_i g_i(\overline{x}) \leq 0$, $\forall (\lambda, \mu) \in \mathbb{R}^p_+ \times \mathbb{R}^q_+ \times \mathbb{R}^{r-q}$, and hence $L(\lambda, \mu, \overline{x}) \leq f_0(\overline{x}) = L(\overline{\lambda}, \overline{\mu}, \overline{x})$, by Kuhn-Tucker Conditions (i), (ii). □

7.2.3 Duality in convex programming

Let $C = \text{dom}(P)$. Consider the function $g : \mathbb{R}^p_+ \times \mathbb{R}^q_+ \times \mathbb{R}^{r-q} \to \mathbb{R} \cup \{-\infty\}$ defined by $g(\lambda, \mu) = \inf_{x\in C}(f_0(x) + \sum_{i\in I}(\lambda_i f_i)(x) + \sum_{i\in J\cup K} \mu_i g_i(x))$. One can easily show that g is concave with respect to (λ, μ). We call *dual program* of (P) the program (P^*):

$$\max_{(\lambda,\mu)\in\mathbb{R}^p_+\times\mathbb{R}^q_+\times\mathbb{R}^{r-q}} g(\lambda, \mu).$$

Theorem 7.2.3 (Duality Theorem).
(i) If Problem (P) has a saddle-point $(\overline{\lambda}, \overline{\mu}, \overline{x})$, then $\overline{x}$ is a solution to (P) and $(\overline{\lambda}, \overline{\mu})$ is a solution to (P^*). Moreover, one has $g(\overline{\lambda}, \overline{\mu}) = f_0(\overline{x})$.
(ii) Conversely, if $\text{dom}(P) = \text{dom}(f_0)$, *if $\overline{x}$ is a solution to (P) and $(\overline{\lambda}, \overline{\mu}) \in \mathbb{R}^p_+ \times \mathbb{R}^q_+ \times \mathbb{R}^{r-q}$ are such that $g(\overline{\lambda}, \overline{\mu}) = f_0(\overline{x})$, then $(\overline{\lambda}, \overline{\mu}, \overline{x})$ is a saddle-point of L.*

Proof. (i) If $(\overline{\lambda}, \overline{\mu}, \overline{x})$ is a saddle-point of L, from Theorem (7.2.2), $\overline{x}$ is a solution to (P) and one has $f_i(\overline{x}) \leq 0$, $\forall i \in I$, $g_i(\overline{x}) \leq 0, \forall i \in J$, $g_i(\overline{x}) = 0$, $\forall i \in I$. Hence, $g(\lambda, \mu) \leq f_0(\overline{x})$, $\forall (\lambda, \mu) \in \mathbb{R}^p_+ \times \mathbb{R}^q_+ \times \mathbb{R}^{r-q}$ and $g(\overline{\lambda}, \overline{\mu}) \leq f_0(\overline{x})$. But from the very definition of the saddle-point, one has $g(\overline{\lambda}, \overline{\mu}) \geq L(\overline{\lambda}, \overline{\mu}, \overline{x}) = f_0(\overline{x})$. In other words, $(\overline{\lambda}, \overline{\mu})$ is a solution to (P^*).

(ii) If $g(\overline{\lambda}, \overline{\mu}) = f_0(\overline{x})$ with $\overline{x}$ solution to (P), one has:

$$f_0(\overline{x}) = g(\overline{\lambda}, \overline{\mu}) \leq f_0(x) + \sum_{i\in I}(\overline{\lambda}_i f_i)(x) + \sum_{i\in J\cup K} \overline{\mu}_i g_i(x), \ \forall x \in C.$$

Taking $x = \overline{x}$, one gets $0 \geq \sum_{i\in I} \overline{\lambda}_i f_i(\overline{x}) + \sum_{i\in J\cup K} \overline{\mu}_i g_i(\overline{x}) \geq 0$. Thus $\overline{\lambda}_i f_i(\overline{x}) = 0$, $\forall i \in I$ and $\overline{\mu}_i g_i(\overline{x}) = 0$, $\forall i \in J$. Since we have $g_i(\overline{x}) = 0, \forall i \in K$, Kuhn-Tucker Conditions (i) and (ii) are fulfilled. On the other hand, we have

$$\begin{aligned} g(\overline{\lambda}, \overline{\mu}) = f_0(\overline{x}) &= f_0(\overline{x}) + \sum_{i\in I} \overline{\lambda}_i f_i(\overline{x}) + \sum_{i\in J\cup K} \overline{\mu}_i g_i(\overline{x}) \\ &\leq f_0(x) + \sum_{i\in I} \overline{\lambda}_i f_i(x) + \sum_{i\in J\cup K} \overline{\mu}_i g_i(x), \ \forall x \in C. \end{aligned}$$

Introduce the function $h : \mathbb{R}^n \to \mathbb{R} \cup \{+\infty\}$ defined by

$$\forall x \in \mathbb{R}^n,\ h(x) = f_0(x) + \sum_{i\in I}(\overline{\lambda}_i f_i)(x) + \sum_{i\in J\cup K} \overline{\mu}_i g_i(x).$$

We have

$$\forall x \in \mathbb{R}^n,\ h(x) + \delta_C(x) \geq h(\overline{x}) + \delta_C(\overline{x}) = h(\overline{x}).$$

Since ri(dom(P) $\neq \emptyset$, by Proposition 6.5.5, one has:

$$0 \in \partial f_0(\overline{x}) + \sum_{i\in I} \overline{\lambda}_i \partial f_i(\overline{x}) + \sum_{i\in J\cup K} \overline{\mu}_i\{\nabla g_i(\overline{x})\} + N_C(\overline{x}).$$

The Kuhn-Tucker conditions (i), (ii) and (iii) are satisfied. From Assertion (ii) of Theorem 7.2.2 $(\overline{\lambda}, \overline{\mu}, \overline{x})$ is a saddle-point of L. □

7.3 Value function

Let $(P_{\alpha,\beta})$ denote the problem:

$$(P_{\alpha,\beta})\qquad \begin{array}{l}\min f(x) \text{ under the constraints}\\ \begin{cases} f_i(x) \leq \alpha_i, \ \forall i \in I\\ g_i(x) = \beta_i, \ \forall i \in K\\ x \in \mathbb{R}^n,\end{cases}\end{array}$$

where f and f_i, for $i \in I$ are convex functions from $\mathbb{R}^n$ into $\mathbb{R} \cup \{+\infty\}$ and g_i, for $i \in K$ are linear. Here f_i may be linear.

The *value function* is the function V defined as follows:

$$V(\alpha, \beta) = \inf\{f(x) \mid x \in \mathbb{R}^n; f_i(x) \leq \alpha_i,\ \forall i \in I;\ g_i(x) = \beta_i,\ \forall i \in J\}$$

where $\alpha = (\alpha_i)_{i\in I} \in \mathbb{R}^p$ and $\beta = (\beta_i)_{i\in K} \in \mathbb{R}^r$, with $p = \text{card}(I)$ and $r = \text{card}(K)$.

Proposition 7.3.1. *(i) V is convex.*
(ii) If α, α' are such that $\alpha_i \leq \alpha'_i$, $\forall i \in I$, then $V(\alpha, \beta) \geq V(\alpha', \beta)$.

Proof. (i) Let

$$E(\alpha, \beta) = \{x \in \mathbb{R}^n \mid f_i(x) \leq \alpha_i,\ \forall i \in I;\ g_i(x) = \beta_i,\ \forall i \in K\}.$$

One can easily check that for $\lambda \in [0, 1]$, we have the inclusion

$$\lambda E(\alpha, \beta) + (1 - \lambda)E(\alpha', \beta') \subset E(\alpha^\lambda, \beta^\lambda),$$

where $\alpha^\lambda = \lambda\alpha + (1 - \lambda)\alpha'$ and $\beta^\lambda = \lambda\beta + (1 - \lambda)\beta'$. Consequently, if $x \in E(\alpha, \beta), x' \in E(\alpha', \beta'), \lambda \in [0, 1]$, then

$$V(\alpha^\lambda, \beta^\lambda) \leq f(\lambda x + (1 - \lambda)x') \leq \lambda f(x) + (1 - \lambda)f(x').$$

Hence $V(\alpha^\lambda, \beta^\lambda) \leq \lambda V(\alpha, \beta) + (1 - \lambda)V(\alpha', \beta')$.

(ii) Since $E(\alpha, \beta) \subset E(\alpha', \beta)$ if $\alpha_i \leq \alpha'_i, \forall i \in I$, one has

$$V(\alpha, \beta) = \inf_{x\in E(\alpha,\beta)} f(x) \geq \inf_{x\in E(\alpha',\beta)} f(x) = V(\alpha', \beta).$$ □

Lemma 7.3.1. *If* $(\lambda, \mu) \in \partial V(\alpha, \beta)$*, then* $\lambda \in \mathbb{R}^n_-$.

Proof. If $(\lambda, \mu) \in \partial V(\alpha, \beta)$ then:

$$\forall(\alpha', \beta') \in \mathbb{R}^p \times \mathbb{R}^r,\ V(\alpha', \beta') - V(\alpha, \beta) \geq \lambda \cdot (\alpha' - \alpha) + \mu \cdot (\beta' - \beta).$$

In particular, for $\alpha' \geq \alpha$ and $\beta' = \beta$, we have $\lambda \cdot (\alpha' - \alpha) \leq 0$. In other words, $\forall h \in \mathbb{R}^p_+$, $\lambda \cdot h \leq 0$. Hence $\lambda_i \leq 0$, $\forall i \in I$. □

Let $\overline{x} \in \mathbb{R}^n$ verify $f_i(\overline{x}) \leq \alpha_i$, $\forall i \in I$; $g_i(\overline{x}) = \beta_i$, $\forall i \in K$. As in the previous section, we called vector of *multipliers* at $\overline{x}$ of Problem $(P_{\alpha,\beta})$ any vector $(\lambda, \mu) \in \mathbb{R}^p \times \mathbb{R}^r$ which satisfies $\lambda \geq 0, \lambda_i(f_i(\overline{x}) - \alpha_i) = 0, \forall i \in I$ and is such that the function

$$\varphi(x) = f(x) + \sum_{i \in I} \lambda_i(f_i(x) - \alpha_i) + \sum_{i \in K} \mu_i(g_i(x) - \beta_i) + \delta_C(x)$$

where $C = \operatorname{dom}(f) \cap (\cap_{i \in I} \operatorname{dom}(f_i))$, is minimal at $\overline{x}$. Recall that, if $\overline{x}$ satisfies $f_i(\overline{x}) \leq \alpha_i$, $\forall i \in I$, $g_i(\overline{x}) = \beta_i, \forall i \in K$ and if $\overline{x}$ has a multiplier for $(P_{\alpha,\beta})$, then $\overline{x}$ is a solution to $(P_{\alpha,\beta})$ (Proposition 7.2.2).

Theorem 7.3.1. *If $\overline{x}$ is a solution to $(P_{\alpha,\beta})$, then the set of multipliers at $\overline{x}$ is equal to $-\partial V(\alpha, \beta)$.*

Proof. Let $(\lambda, \mu) \in -\partial V(\alpha, \beta)$. From lemma 7.3.1, $\lambda \geq 0$. Since $\overline{x}$ is a solution, $V(\alpha, \beta) = f(\overline{x})$. On the other hand, if $x \in E(\alpha, \beta)$ then $f(x) \geq V(\alpha, \beta)$. We also have :

$$\forall(\alpha', \beta') \in \mathbb{R}^p \times \mathbb{R}^r,\ V(\alpha', \beta') - V(\alpha, \beta) \geq -\lambda \cdot (\alpha' - \alpha) - \mu \cdot (\beta' - \beta)$$

or :

$$V(\alpha', \beta') + \lambda \cdot (\alpha' - \alpha) + \mu \cdot (\beta' - \beta) \geq V(\alpha, \beta).$$

Let $x \in C$. Define (α', β') by $\alpha'_i = f_i(x)$, $\forall i \in I$, $\beta'_i = g_i(x)$, $\forall i \in K$. Obviously, $x \in E(\alpha', \beta')$ and hence $V(\alpha', \beta') \leq f(x)$. Therefore:

$$\forall x \in C,\ f(x) + \sum_{i \in I} \lambda_i(f_i(x) - \alpha_i) + \sum_{i \in K} \mu_i(g_i(x) - \beta_i) \geq V(\alpha, \beta) \tag{7.2}$$

Since $V(\alpha, \beta) = f(\overline{x}) \geq f(\overline{x}) + \sum_{i \in I} \lambda_i(f_i(\overline{x}) - \alpha_i) + \sum_{i \in K} \mu_i(g_i(\overline{x}) - \beta_i)$, we see that $\overline{x}$ minimizes φ. Replacing x by $\overline{x}$ in Relation 7.2, we obtain $\sum_{i \in I} \lambda_i(f_i(\overline{x}) - \alpha_i) \geq 0$. Thus $\sum_{i \in I} \lambda_i(f_i(\overline{x}) - \alpha_i) = 0$ which implies $\lambda_i(f_i(\overline{x}) - \alpha_i) = 0$, $\forall i \in I$. We have proved that (λ, μ) is a multiplier at $\overline{x}$ for $(P_{\alpha,\beta})$.

Let us prove the converse. Let (λ, μ) be a multiplier at $\overline{x}$ for $(P_{\alpha,\beta})$. We have:

$$\forall x \in C,\ f(x) + \sum_{i \in I} \lambda_i(f_i(x) - \alpha_i) + \sum_{i \in K} \mu_i(g_i(x) - \beta_i)$$

$$\geq f(\overline{x}) + \sum_{i \in I} \lambda_i(f_i(\overline{x}) - \alpha_i) + \sum_{i \in J \cup K} \mu_i(g_i(\overline{x}) - \beta_i) = V(\alpha, \beta).$$

Let $x \in E(\alpha', \beta')$. Then $x \in \text{dom}(f_i)$, $\forall i \in I$. Either $x \in \text{dom}(f)$ (and hence $x \in C$), or $f(x) = +\infty$, in both cases, we have: $f(x) + \sum_{i \in I} \lambda_i(\alpha'_i - \alpha_i) + \sum_{i \in K} \mu_i(\beta'_i - \beta_i) \geq f(\overline{x}) + \sum_{i \in I} \lambda_i(f_i(\overline{x}) - \alpha_i) + \sum_{i \in K} \mu_i(g_i(\overline{x}) - \beta_i) \geq V(\alpha, \beta)$.
Hence: $f(x) \geq V(\alpha, \beta) - \lambda \cdot (\alpha' - \alpha) - \mu \cdot (\beta' - \beta)$ which implies $V(\alpha', \beta') \geq V(\alpha, \beta) - \lambda \cdot (\alpha' - \alpha) - \mu \cdot (\beta' - \beta)$. In other words, $(-\lambda, -\mu) \in \partial V(\alpha, \beta)$. □

Corollary 7.3.1. *Let $\overline{x}$ be a solution to $(P_{\alpha,\beta})$. The function V is differentiable in (α, β) if, and only if, there exists a unique multiplier (λ, μ) at $\overline{x}$. In this case, $\nabla V(\alpha, \beta) = (-\lambda, -\mu)$.*

Proof. It follows from Proposition 6.6.2. □

Exercises

Exercise 7.1. Demonstrate Lemma 7.2.2.

Exercise 7.2. Let f be a convex function from $\mathbb{R}^n$ into $\mathbb{R}$. Consider the problem

$$(P) \qquad \min_{x \in U} f(x)$$

where U is nonempty convex set of $\mathbb{R}^n$. If $\overline{x}$ is a solution to (P), we say that $\overline{x}$ is a global solution.
(i) Show that the set of solutions to (P) is convex.
(ii) Recall that $\overline{x}$ is a local solution to (P) if there exists an open ball B centered at $\overline{x}$ such that $\forall x \in B \cap U$, $f(x) \geq f(\overline{x})$. Show that any local solution is global solution.

Exercise 7.3. Let f be the identity function on $\mathbb{R}$, i.e. $f(x) = x$, $\forall x \in \mathbb{R}$.
(i) Solve Problem (P): $\max\{f(x) \mid x^2 \leq 0\}$.
(ii) Has this problem a saddle-point? Why?

Exercise 7.4. Consider the following problem (P):

$$(P) \qquad \begin{array}{c} \min(x^2 + y^2) \text{ under the constraint} \\ 2x + y \leq -4. \end{array}$$

(i) Solve (P).
(ii) Write its dual program (P^*).
(iii) Solve (P^*).
(iv) Check the duality theorem.

Exercise 7.5.

$$\begin{array}{c} \min(x + y) \text{ under the constraint} \\ x^2 + y^2 \leq 1. \end{array}$$

Exercise 7.6. Let E be the ellipsoid of $\mathbb{R}^3$, the equation of which is: $x^2+y^2+z^2/2 \leq 1$, and let Δ be the hyperplane, the equation of which is $x+y+z=3$.
(i) Find the points $M = (x_1^M, x_2^M, x_3^M)$, $M' = (x_1^{M'}, x_2^{M'}, x_3^{M'})$ of E and Δ which minimize the euclidean distance between E and Δ.
(ii) Check that the vector $(x_1^{M'} - x_1^M, x_2^{M'} - x_2^M, x_3^{M'} - x_3^M)$ is in the normal cone at (x_1^M, x_2^M, x_3^M) to E and orthogonal to Δ.

Exercise 7.7. Recall that if $A \subset \mathbb{R}^n$, the polar set of A, A^0, is defined as follows:

$$A^0 = \{y \in \mathbb{R}^n \mid x \cdot y \leq 1, \forall x \in A\}.$$

Consider the support function δ_A^* of A (i.e. $\delta_A^*(y) = \sup\{x \cdot y \mid x \in A\}$). Notice that $\delta_A^*(y) = \inf\{-x \cdot y \mid x \in A\}$. Therefore, $A^0 = \{y \mid \delta_A^*(y) \leq 1\}$.
(i) Use Proposition 7.2.1 to compute the functions δ_A^* in the following cases
(a) $A = \{x \in \mathbb{R}^n \mid x_k \geq 0, \forall k = 1, \dots, n; \sum_{k=1}^n x_k \leq 1\}$.
(b) $A = \{x \in \mathbb{R}^2 \mid (x_1-1)^2 + x_2^2 \leq 1\}$.
(c) $A = \{x \in \mathbb{R}^2 \mid x_1 \leq 1 - \sqrt{1+x_2^2}\}$.
(d) $A = \{x \in \mathbb{R}^n \mid \sum_{i=1}^n |x_i| \leq 1\}$.
(ii) Find the polar sets of these A's.

Exercise 7.8. Let $p \in \mathbb{R}$ be given. Solve

$$\min(x^2 + (y-p)^2) \text{ under the constraints } \begin{cases} x^2 + y^2 \leq 1 \\ x \leq y \end{cases}$$

Give a geometric interpretation of the results.

Exercise 7.9. (i) Solve:

$$\min(x^2 + y^2 + (z-2)^2) \text{ under the constraints } \begin{cases} \dfrac{x^2}{a^2} + \dfrac{y^2}{(b^2+z^2)} \leq 1 \\ y - p = 0 \end{cases}$$

where $a > 0, b > 0$ and $-b < p < b$.
(ii) Check the duality theorem.

Exercise 7.10. Consider Problem $(P_{\alpha,\beta})$ in Section 7.3 with functions f and $(f_i)_{i \in I}$ convex, differentiable from $\mathbb{R}^n$ into $\mathbb{R}$. Let V be the value function of $(P_{\alpha,\beta})$. Show that, if $\overline{x}$ is a solution to $(P_{\alpha,\beta})$, and if the gradients $(\nabla f_i(\overline{x}))_{i \in I}, (\nabla g_i(\overline{x}))_{i \in K}$ are linearly independent, then V is differentiable at (α, β).

Exercise 7.11 (Farkas Lemma). Let $f_1, \dots, f_p, g$ be linear forms on $\mathbb{R}^n$. Noticing that the following claim

$$f_k(x) \geq 0, \ \forall k = 1, \dots, p \Rightarrow g(x) \geq 0$$

is equivalent to say that 0 solves

$$\begin{aligned} &\min g(x) \text{ under the constraints} \\ &f_k(x) \geq 0, \ \forall k = 1, \ldots, p \end{aligned}$$

Use corollary 7.2.1 to show that it is equivalent to $\exists \lambda_k \geq 0$, $\forall k = 1, \ldots, p$ and $g = \sum_{k=1}^{p} \lambda_k f_k$.

Exercise 7.12 (Linear Programming). Let $c \in \mathbb{R}^n, b \in \mathbb{R}^m$ and A be a $(m \times n)$ real matrix. Prove that $\overline{x}$ solves the problem

$$\min c \cdot x \text{ under the constraints } \begin{cases} Ax \geq b \\ x \geq 0 \end{cases}$$

if, and only if, there exists $p \in \mathbb{R}^m_+, q \in \mathbb{R}^n_+$, such that ${}^tc = {}^tp{\cdot}A + {}^tq$, $(A\overline{x} - b) \geq 0, \overline{x} \geq 0, p \cdot (A\overline{x} - b) = 0, p \cdot A\overline{x} = c \cdot \overline{x}$, where ${}^tp, {}^tq, {}^tc$ are the transposed of p, q, c.

Exercise 7.13. Consider the function $u : \mathbb{R}^2_+ \to \mathbb{R}_+$ defined by $u(x_1, x_2) = x_1^\alpha x_2^\beta$, where $(\alpha, \beta) \in \mathbb{R}^2_{++}$ and $\alpha + \beta = 1$.
a) Show that u is concave and continuous on $\mathbb{R}^2_+$ and differentiable on $\mathbb{R}^2_{++}$.
b) For $\omega \in \mathbb{R}^2_+ \setminus \{0\}$ and $p \in \mathbb{R}^2_{++}$, we want to solve the following problem:

$$(P) \qquad \text{Maximize } u(x_1, x_2) \text{ under the constraints } \begin{cases} p_1 x_1 + p_2 x_2 \leq p \cdot \omega \\ x_1 \geq 0 \\ x_2 \geq 0. \end{cases}$$

i) Prove that this problem has a solution.
ii) Show that this solution is in $\mathbb{R}^2_{++}$.
iii) Find a solution (x_1, x_2, λ) to the following system:

$$\begin{cases} \alpha x_1^{(\alpha-1)} x_2^\beta = \lambda p_1 \\ \beta x_1^\alpha x_2^{(\beta-1)} = \lambda p_2 \\ p_1 x_1 + p_2 x_2 = p \cdot \omega \\ x_1 > 0, x_2 > 0, \lambda > 0. \end{cases}$$

iv) Show that this solution (x_1, x_2, λ) verifies Kuhn-Tucker's Conditions of Problem (P), and hence (x_1, x_2) solves Problem (P).

8. Non Convex Optimization

In the previous chapter, we study the minimization problem of a convex function under convex constraints. In this chapter, we study the minimization problem for two other classes of functions. The first ones are quasi-convex functions. We know (see Chapter 5) that for a convex function f, any level set of f is convex. But the converse is not true. The functions which have the property that every level set is convex are precisely called *quasi-convex* functions. In the first section we give necessary and sufficient conditions for a point to be a solution to a minimization problem of a quasi-convex function under *convex* constraints.
In the second section of the present chapter, we study the minimization problem of a continuously differentiable function, where the constraints are defined a finite number of inequalities on continuously differentiable functions. The function to be minimized and the function which define the constraints are not necessarily convex. We need for the proofs, the Local inversion theorem, the proof of which is given in Appendix.

8.1 Quasi-convex functions

Let $f : U \to \mathbb{R}$ be a function where U is a non-empty convex subset of $\mathbb{R}^n$. We say that f is *quasi-convex* if, $\forall \alpha \in \mathbb{R}$, the level set $\text{lev}_{\leq\alpha}(f) = \{x \in U \mid f(x) \leq \alpha\}$ is convex.
Recall that a real function from an interval $I \subset \mathbb{R}$ is non-decreasing if $\forall x \in I, \forall y \in I$, $x \leq y \Rightarrow f(x) \leq f(y)$. It is non-increasing if $\forall x \in I, \forall y \in I$, $x \leq y \Rightarrow f(x) \geq f(y)$. It is *monotonic* if either it is non-decreasing or it is non-increasing.

Proposition 8.1.1. *(i) A monotonic continuous function from an interval I into $\mathbb{R}$ is quasi-convex.*
(ii) If f is continuous and if there exists $t_0 \in I$ such that f is non-increasing for $x \leq t_0$ and non-decreasing for $x \geq t_0$ then f is quasi-convex from I into $\mathbb{R}$.
(iii) If for $i \in I$, the function f_i is quasi-convex, then $\sup\{f_i \mid i \in I\}$ is quasi-convex.
(iv) Let f be a function from a convex, nonempty set U of $\mathbb{R}^n$ into $\mathbb{R}$. The

function f is quasi-convex if, and only if, $\forall x \in U, \forall y \in U$, the function $g : [0,1] \to \mathbb{R}$ defined by $g(\lambda) = f(\lambda x + (1-\lambda)y)$ is quasi-convex.
(v) If f is quasi-convex, if g is non-decreasing from $\mathbb{R}$ into $\mathbb{R}$, then $g \circ f$ is quasi-convex.

Proof. (i) Assume f non-decreasing on I. Let $\alpha \in \mathbb{R}$. If $\mathrm{lev}_{\leq\alpha}(f)$ is empty, then it is convex. Assume $\mathrm{lev}_{\leq\alpha}(f)$ nonempty and let $x, y \in \mathrm{lev}_{\leq\alpha}(f)$ such that $x \leq y$. Let $\lambda \in]0,1[$. Then $\lambda x + (1-\lambda)y \leq y$ which implies $f(\lambda x + (1-\lambda)y) \leq f(y) \leq \alpha$. Hence $\lambda x + (1-\lambda)y \in \mathrm{lev}_{\leq\alpha}(f)$. Use the same method when f is non-increasing.
(ii) Let $x \in \mathrm{lev}_{\leq\alpha}(f), y \in \mathrm{lev}_{\leq\alpha}(f)$ such that $x \leq y$ and let $\lambda \in]0,1[$.
If $x \leq t_0, y \leq t_0$, then $\lambda x+(1-\lambda)y \geq x$ and hence $f(\lambda x+(1-\lambda)y) \leq f(x) \leq \alpha$, i.e. $\lambda x + (1-\lambda)y \in \mathrm{lev}_{\leq\alpha}(f)$.
By the same argument, the same conclusion holds if $x \geq t_0, y \geq t_0$.
Assume $x \leq t_0 \leq y$. Then we have $x \leq \lambda x + (1-\lambda)y \leq y$. One can check that $f(\lambda x+(1-\lambda)y) \leq \max\{f(x), f(y)\} \leq \alpha$. In other words $\lambda x+(1-\lambda)y \in \mathrm{lev}_{\leq\alpha}(f)$.
(iii) The proof is easy.
(iv) Assume f quasi-convex. Let $x, y \in U$ and $\alpha \in \mathbb{R}$. We will show that the set $\mathrm{lev}_{\leq\alpha}(g) = \{\lambda \in [0,1] \mid g(\lambda) \leq \alpha\}$ is convex. Let $\lambda, \lambda' \in \mathrm{lev}_{\leq\alpha}(g)$. Then $\lambda x+(1-\lambda)y \in \mathrm{lev}_{\leq\alpha}(f), \lambda' x+(1-\lambda')y \in \mathrm{lev}_{\leq\alpha}(f)$. Since $\mathrm{lev}_{\leq\alpha}(f)$ is convex, $\mu(\lambda x+(1-\lambda)y)+(1-\mu)(\lambda' x+(1-\lambda')y) \in \mathrm{lev}_{\leq\alpha}(f)$ for all $\mu \in [0,1]$. Hence $\mu\lambda + (1-\mu)\lambda' \in \mathrm{lev}_{\leq\alpha}(g)$.
Conversely, let $x, y \in \mathrm{lev}_{\leq\alpha}(f)$. Assume g quasi-convex. Let $\alpha \in \mathbb{R}$. We prove that $\mathrm{lev}_{\leq\alpha}(f)$ is convex. We have $g(1) \leq \alpha, g(0) \leq \alpha$. Let $\lambda \in [0,1]$. Then $g(\lambda) = f(\lambda x + (1-\lambda)y) \leq \alpha$. That means that $\lambda x + (1-\lambda)y \in \mathrm{lev}_{\leq\alpha}(f)$.
(v) The proof is obvious. □

Proposition 8.1.2. *Let f be a function from a convex, nonempty set U of $\mathbb{R}^n$ into $\mathbb{R}$. f is quasi-convex if, and only if:*

$$\forall x \in U, \forall y \in U, \forall \lambda \in [0,1],\ f(\lambda x + (1-\lambda)y) \leq \max(f(x), f(y)) \tag{8.1}$$

Proof. Assume f quasi-convex. We will show that f verifies Inequality (8.1). Define $\alpha = \max(f(x), f(y))$. Then x and y belong to $\mathrm{lev}_{\leq\alpha}(f)$. If f is quasi-convex then $\mathrm{lev}_{\leq\alpha}(f)$ is convex. Thus $\lambda x + (1-\lambda)y \in \mathrm{lev}_{\leq\alpha}(f)$ and $f(\lambda x + (1-\lambda)y) \leq \alpha$.
Conversely, assume that f satisfies Condition (8.1). We will prove that f is quasi-convex. Let α be a real number. If x and y belong to $\mathrm{lev}_{\leq\alpha}(f)$ and $\lambda \in [0,1]$, then $f(\lambda x + (1-\lambda)y) \leq \max(f(x), f(y)) \leq \alpha$. That means that $\lambda x + (1-\lambda)y \in \mathrm{lev}_{\leq\alpha}(f)$.
The proof is complete. □

The previous proposition allows us to strengthen the notion of quasi-convexity. A function $f : U \to \mathbb{R}$, where U is a non-empty convex subset of $\mathbb{R}^n$, is said to be *strictly quasi-convex* if it is quasi-convex and if $\forall x \in U, \forall y \in U$,

$f(x) < f(y) \Rightarrow \forall \lambda \in]0,1[\,,\ f(\lambda x + (1-\lambda)y) < f(y)$.
Any convex function is strictly quasi-convex (see Exercise 8.3).
A function $f : U \longrightarrow \mathbb{R}$, where U is a non-empty convex subset of $\mathbb{R}^n$, is said to be *strongly strictly quasi-convex* if it is quasi-convex and if $\forall x \in U, \forall y \in U$, $x \neq y \Rightarrow \forall \lambda \in]0,1[, f(\lambda x + (1-\lambda)y) < \max\{f(x), f(y)\}$.
Any strongly strictly convex function is strictly quasi-convex.

As we will show by an example, we would like to mention that for a continuous quasi-convex function $\mathbb{R}^n$ into $\mathbb{R}$, the *closure* of the set $\text{lev}_{<\alpha}(f) = \{x \in \mathbb{R}^n \mid f(x) < \alpha\}$, where $\alpha \in \mathbb{R}$ may be different of the level set $\text{lev}_{\leq\alpha}(f)$: Consider the following continuous quasi-convex function in $\mathbb{R}$ defined by

$$f(x) = \begin{cases} x & \text{if } x \leq 0; \\ 0 & \text{if } x \in [0,1] \\ x-1 & \text{if } x \geq 1. \end{cases}$$

Note that it is not strictly quasi-convex ($0 \in]-1/2, 1/2[$ but $f(0) = f(1/2) > f(-1/2)$). Moreover, since $\text{lev}_{<0}(f) =]-\infty, 0[$, it follows $\overline{\text{lev}_{<0}(f)} =]-\infty, 0]$, but $\text{lev}_{\leq 0}(f) =]-\infty, 1]$.
But if a function f is continuous, *strictly* quasi-convex then we have the property that $\forall \alpha \in \mathbb{R}$, if $\text{lev}_{<\alpha}(f) \neq \emptyset$ then $\overline{\text{lev}_{<\alpha}(f)} = \text{lev}_{\leq\alpha}(f)$ (see Exercise 8.3).

Proposition 8.1.3. *Let f be a real, continuous function on an nonempty interval I of $\mathbb{R}$. Then f is quasi-convex if, and only if, either it is monotonic on I, or there exists $t_0 \in I$ such that f is non-increasing for $t \leq t_0$, and non-decreasing for $t \geq t_0$.*

Proof. If f is monotonic, or non-increasing for $t \leq t_0$, and non-decreasing for $t \geq t_0$, then from Proposition 8.1.1, Assertions (i) and (ii), it is quasi-convex.

Conversely, let f be quasi-convex. We will prove that f satisfies the properties given in the statement of the present proposition. Let $\alpha = \inf\{f(t) \mid t \in I\}$, we will distinguish several cases.
First, assume $I = [a, b]$.
Let t_0 satisfy $f(t_0) = \alpha$. Let t, t' be such that $a \leq t \leq t' \leq t_0$. By Proposition 8.1.2, we have $f(a) \geq f(t) \geq f(t_0)$, and hence $f(t) \geq f(t')$ by the same proposition. That means that f is non-increasing in $[a, t_0]$. We prove by the same way, that f is non-decreasing in $[t_0, b]$.
Assume now that $I = [a, b[$ with $b \leq +\infty$.
There are two sub-cases: (i) $f(t_0) = \alpha$, for some $t_0 \in [a, b[$; (ii) $f(t) > \alpha$, $\forall t \in [a, b[$.
Consider case (i). Let $a \leq t \leq t' \leq t_0$. By Proposition 8.1.2, we have $f(a) \geq f(t) \geq f(t_0)$, and hence $f(t) \geq f(t')$. That means that f is non-increasing in $[a, t_0]$. We prove by the same way, that f is non-decreasing in $[t_0, b[$.
Consider case (ii). We will prove that f is non-increasing. Let $(t'_n)_{n\in\mathbb{N}}$ be a

sequence such that $\lim_{n\to+\infty} f(t'_n) = \alpha$. Since f is continuous and since the infimum is not attained, we may assume that $(t'_n)_{n\in\mathbb{N}}$ is increasing and that $t'_n \to b$. Define a sequence $(t_n)_{n\in\mathbb{N}}$ as follows:

$$f(t_0) = \min\{f(t) \mid t \in [a, t'_0]\},$$

and for $n \geq 1$:

$$f(t_n) = \min\{f(t) \mid t \in [t_{n-1}, t'_n]\}.$$

Such a sequence exists since f is continuous. We observe that :
(a) the sequence $(f(t_n))_{n\in\mathbb{N}}$ is non-increasing,
(b) $f(t_n) \leq f(t'_n)$, $\forall n \in \mathbb{N}$, and hence $f(t_n) \to \alpha$,
(c) $(t_n)_{n\in\mathbb{N}}$ is non-decreasing and $t_n \to b$ when $n \to +\infty$,
(d) $f(t_{n-1}) \geq f(t) \geq f(t_n)$, $\forall t \in [t_{n-1}, t_n]$ by (a), Proposition 8.1.2 and the definition of t_n.
Now, let $a \leq t \leq t'$. If $t_n \leq t \leq t' \leq t_{n+1}$, from (d) one gets $f(t') \leq f(t)$ by Proposition 8.1.2. If $t_{n-1} \leq t \leq t_n$ and $t_{m-1} \leq t' \leq t_m$ with $m > n$, then $f(t') \leq f(t)$ from (d) and (a). We have proved that f is non-increasing.
Assume that $I =]a, b]$ with $a \geq -\infty$.
We prove by the same method that, either there exists $t_0 \in]a, b]$ such that f is non-increasing for $t \in]a, t_0]$ and non-decreasing for $t \in [t_0, b]$, or f is non-decreasing on $]a, b]$.
Finally, consider the case where $I =]a, b[$ with $a \geq -\infty$ and $b \leq +\infty$.
We consider two cases : (i) $f(t_0) = \alpha$, for some $t_0 \in]a, b[$, (ii) $f(t) > \alpha$, $\forall t \in]a, b[$. In case (i), by the same method, we prove that f is non-increasing for $t \in]a, t_0]$ and non-decreasing in $[t_0, b[$.
In case (ii), let $(t'_n)_{n\in\mathbb{N}}$ be a sequence such that $\lim f(t'_n) = \alpha$. Since f is continuous, either $(t'_n)_{n\in\mathbb{N}}$ converges to b or it converges to a.
Assume that $(t'_n)_{n\in\mathbb{N}}$ converges to b. Let t, t' be such that $t \leq t'$. Let $a < c < t$. We can easily prove, using the same technique as before, that f is non-increasing in $[c, b[$. Hence $f(t) \geq f(t')$. Since t is arbitrary in $]a, b[$, we have actually proved that f is non-increasing in $]a, b[$. Similarly, if the sequence $(t'_n)_{n\in\mathbb{N}}$ converges to a then f is non-decreasing in $]a, b[$. The proof is now complete. □

Proposition 8.1.4. *Let f be a differentiable function from an open, convex, non-empty set U of $\mathbb{R}^n$ into $\mathbb{R}$. f is quasi-convex if, and only if:*

$$\forall x \in U, \forall y \in U, \ f(y) \leq f(x) \Rightarrow \nabla f(x) \cdot (y - x) \leq 0 \tag{8.2}$$

Proof. Assume that f is quasi-convex and $f(y) \leq f(x)$. From Proposition 8.1.2, we have:

$$\forall \lambda \in [0, 1], \ f((1-\lambda)x + \lambda y) \leq \max(f(x), f(y)) = f(x).$$

Hence : $(f(x + \lambda(y - x)) - f(x))/\lambda \leq 0, \ \forall \lambda \in]0, 1[$.
Thus:

$$\nabla f(x) \cdot (y - x) = \lim_{\lambda \to 0, \lambda > 0} \frac{f(x + \lambda(y - x)) - f(x)}{\lambda} \leq 0.$$

Conversely, let us consider a function f verifying Condition (8.2). Define: $\varphi(\lambda) = f(y + \lambda(x - y))$ for $\lambda \in [0, 1]$ and $f(x) \geq f(y)$. By Proposition 8.1.2, to show φ is quasi-convex is to show that $\varphi(\lambda) \leq \varphi(1), \forall \lambda \in [0, 1]$.
Assume the contrary. There exists $\overline{\lambda}$ which verifies $\varphi(\overline{\lambda}) > \varphi(1) \geq \varphi(0)$. By the Mean value theorem (Theorem A.2.2, in Appendix), there exists λ_1, λ_2 such that $0 < \lambda_1 < \overline{\lambda} < \lambda_2 \leq 1$ and $\varphi'(\lambda_1) > 0, \varphi'(\lambda_2) < 0$.
If $\varphi(\lambda_1) \leq \varphi(\lambda_2)$ then $\varphi'(\lambda_2)(\lambda_1 - \lambda_2) \leq 0$. That is a contradiction with Inequality (8.2) because $\varphi'(\lambda_2)(\lambda_1 - \lambda_2) > 0$.
If $\varphi(\lambda_1) \geq \varphi(\lambda_2)$ then $\varphi'(\lambda_1)(\lambda_2 - \lambda_1) \leq 0$. Again, we have a contradiction since $\varphi'(\lambda_1)(\lambda_2 - \lambda_1) > 0$.
Summing up, we have $\varphi(\lambda) \leq \varphi(1), \forall \lambda \in [0, 1]$. From Proposition 8.1.1, Assertion (iv), f is quasi-convex. □

In the following, for a twice differentiable function f from $\mathbb{R}^n$ into $\mathbb{R}$, to simplify the notation, we write $f''(x)$ instead of $((\partial^2 f / \partial x_i \partial x_j)(x))_{\substack{i=1,\ldots,n,\\ j=1,\ldots,n}}$.

Proposition 8.1.5. *Let f be a twice differentiable, quasi-convex function from $\mathbb{R}^n$ into $\mathbb{R}$. Let $\alpha \in \mathbb{R}$ and let $\overline{x}$ verify $f(\overline{x}) = \alpha$. Let $H_\alpha = \{x \in \mathbb{R}^n \mid \nabla f(\overline{x}) \cdot (x - \overline{x}) = 0\}$. Then:*

$$\forall x \in H_\alpha, \ (x - \overline{x}) \cdot (f''(\overline{x})(x - \overline{x})) \geq 0.$$

Proof. Define the function g on $\mathbb{R}$ by $g(t) = f(\overline{x} + t(x - \overline{x}))$ with $x \in H_\alpha$. Then g is quasi-convex (see Assertion (iv) of Proposition 8.1.1). We have:

$$g'(0) = \nabla f(\overline{x}) \cdot (x - \overline{x}) = 0,$$

$$g'(t) = t(x - \overline{x}) \cdot (f''(\overline{x})(x - \overline{x})) + t\varepsilon(t) \text{ where } \varepsilon(t) \to 0 \text{ when } t \to 0.$$

Assume $(x - \overline{x}) \cdot (f''(\overline{x})(x - \overline{x})) < 0$. There then exists $t_1 < 0, t_2 > 0$, such that $t \in [t_1, 0[\Rightarrow g'(t) > 0$, and $t \in]0, t_2] \Rightarrow g'(t) < 0$. Choose t_1, t_2 such that $g(t_1) = g(t_2) < g(0)$. Then the set $\{t \mid g(t) \leq g(t_1)\}$ is in $]-\infty, t_1] \cup [t_2, +\infty[$ which is not convex. That is a contradiction since g is quasi-convex. Hence, $(x - \overline{x}) \cdot (f''(\overline{x})(x - \overline{x})) \geq 0$. □

We have seen (Exercise 5.14) that if f is convex, twice differentiable on $\mathbb{R}^n$, then for all $x \in \mathbb{R}^n$, $f''(x)$ defines a nonnegative quadratic form over $\mathbb{R}^n$. Here, we have a weaker result for a larger class: the set of quasi-convex functions. The matrix $f''(x)$ defines a nonnegative quadratic form on the hyperplane tangent at x to the closed, convex set $\{y \in \mathbb{R}^n \mid f(y) \leq f(x)\}$.

Proposition 8.1.6. *Let f be a twice differentiable function from $\mathbb{R}^n$ into $\mathbb{R}$. We assume that:*

$$\forall x \in \mathbb{R}^n, \forall h \in \mathbb{R}^n \setminus \{0\}, \ \nabla f(x) \cdot h = 0 \Rightarrow h \cdot (f''(x)h) > 0.$$

Then f is quasi-convex.

Proof. Let x, y in $\mathbb{R}^n$ satisfy $f(y) \leq f(x)$. From Proposition 8.1.4, it suffices to show that $\nabla f(x) \cdot (y-x) \leq 0$. Assume the contrary: $\nabla f(x) \cdot (y-x) > 0$. Define the real function g by $g(t) = f(x + t(y-x)), \forall t \in [0,1]$. We have: $g'(t) = \nabla f(x + t(y-x)) \cdot (y-x)$. Then $g'(0) > 0$. The function g is strictly increasing in a neighborhood of 0. Since g is continuous, it has a maximum on $[0,1]$ at some point t_1 which verifies $g(t_1) > g(0) = f(x) \geq g(1) = f(y)$. Therefore, t_1 is in $]0,1[$. We then have $g'(t_1) = 0, g''(t_1) \leq 0$. But $g'(t_1) = \nabla f(x + t_1(y-x)) \cdot (y-x), g''(t_1) = (y-x) \cdot (f''(x + t_1(y-x))(y-x))$. We have a contradiction with the assumption stated in the proposition. Hence, $\nabla f(x) \cdot (y-x) \leq 0$. □

8.2 Minimization of quasi-convex functions

Let f be a quasi-convex function from an open, convex, nonempty set U of $\mathbb{R}^n$ into $\mathbb{R}$. We assume that f is continuously differentiable.

Consider the following problem (P):

$$(P) \qquad \min_{x \in U} f(x) \text{ under the constraints } \begin{cases} f_i(x) \leq 0, & \forall i \in I \\ g_i(x) \leq 0, & \forall i \in J \end{cases}$$

where for all $i \in I$, f_i is a convex, non-affine function from $\mathbb{R}^n$ into $\mathbb{R}$ and for all $i \in J$, g_i is affine.

Lemma 8.2.1. *Let C, C' be two nonempty convex sets of $\mathbb{R}^n$ and let $\overline{x} \in C \cap C'$. One assumes that $\mathrm{i}(C) \neq \emptyset$.*
If $\mathrm{i}(C) \cap C' = \emptyset$ then there exists $p \neq 0$ such that $p \in N_C(\overline{x}) \cap -N_{C'}(\overline{x})$ where $N_C(\overline{x}), N_{C'}(\overline{x})$ are the normal cones to C, C' at $\overline{x}$.

Proof. From the First Separation Theorem 2.1.7, there exists $p \in \mathbb{R}^n \setminus \{0\}$ such that $p \cdot (x - \overline{x}) \leq p \cdot (x' - \overline{x}), \forall x \in \mathrm{i}(C), \forall x' \in C'$. By continuity, one gets $p \cdot (x - \overline{x}) \leq p \cdot (x' - \overline{x}), \forall x \in C, \forall x' \in C'$. Since $\overline{x} \in C \cap C'$, we have $p \cdot (x - \overline{x}) \leq 0, \forall x \in C$ and $-p \cdot (x' - \overline{x}) \leq 0, \forall x' \in C'$. In other words, $p \in N_C(\overline{x}) \cap -N_{C'}(\overline{x})$. □

Lemma 8.2.2. *Let $f : U \to \mathbb{R}$ be a differentiable quasi-convex function, where U is nonempty convex subset of $\mathbb{R}^n$. Consider the level set $\mathrm{lev}_{\leq\alpha}(f) = \{x \in U \mid f(x) \leq \alpha\}$ for some $\alpha \in \mathbb{R}$. If $\overline{x}$ satisfies $f(\overline{x}) = \alpha, \nabla f(\overline{x}) \neq 0$ then the normal cone to $\mathrm{lev}_{\leq\alpha}(f)$ at $\overline{x}$, $N_{\mathrm{lev}_{\leq\alpha}(f)}(\overline{x})$ is equal to $\{\lambda \nabla f(\overline{x}) \mid \lambda \geq 0\}$. Moreover, $\forall x \in U$, $\nabla f(\overline{x}) \cdot (x - \overline{x}) \geq 0 \Rightarrow f(x) \geq f(\overline{x})$.*

Proof. From Proposition 8.1.4, if $x \in \mathrm{lev}_{\leq\alpha}(f)$ then we have $\nabla f(\overline{x}) \cdot (x - \overline{x}) \leq 0$. In other words, $\nabla f(\overline{x}) \in N_{\mathrm{lev}_{\leq\alpha}(f)}(\overline{x})$.

Let us prove the converse. Take $v \in N_{\mathrm{lev}_{\leq\alpha}(f)}(\overline{x}) \setminus \{0\}$. We first show that $\forall x \in U, v \cdot x \geq v \cdot \overline{x} \Rightarrow f(x) \geq f(\overline{x})$. Assume the contrary. Let x in U verify

$v \cdot x \geq v \cdot \overline{x}$ and $f(x) < f(\overline{x})$. Let $h \in \mathbb{R}^n$ satisfy $v \cdot h > 0$. and $f(x+h) < f(\overline{x})$. We then have a contradiction because $x + h \in \text{lev}_{\leq\alpha}(f), v \in N_{\text{lev}_{\leq\alpha}(f)}(\overline{x}) \Rightarrow v \cdot (x+h) - v \cdot \overline{x} \leq 0$.
Now, if $\{\nabla f(\overline{x})\} \cap \{\mu v \mid \mu \geq 0\} = \emptyset$ then, from the Second Separation theorem 2.1.8, there exists $p \neq 0$ such that $p \cdot \nabla f(\overline{x}) < 0 \leq \mu p \cdot v$, $\forall \mu \geq 0$. Since

$$\nabla f(\overline{x}) \cdot p = \lim_{\mu > 0, \mu \to 0} \frac{f(\overline{x} + \mu p) - f(\overline{x})}{\mu},$$

one has $f(\overline{x} + \mu p) < f(\overline{x})$ for $\mu > 0$, sufficiently small. But we get a contradiction to the previous result since $v \cdot (\overline{x} + \mu p) - v \cdot \overline{x} = \mu p \cdot v \geq 0$. Therefore, $v \in \{\lambda \nabla f(\overline{x}) \mid \lambda \geq 0\}$.
To prove the last statement, observe we have shown that $\forall x \in U$, $v \cdot x \geq v \cdot \overline{x} \Rightarrow f(x) \geq f(\overline{x})$ if $v \in N_{\text{lev}_{\leq\alpha}(f)}(\overline{x}) \backslash \{0\}$. Since $\nabla f(\overline{x}) \in N_{\text{lev}_{\leq\alpha}(f)}(\overline{x}) \backslash \{0\}$, the conclusion then follows. □

As in the previous chapter, the *Slater condition* for Problem (P) is :

$$\exists x_0 \in U, \begin{cases} f_i(x_0) < 0, \ \forall i \in I \\ g_i(x_0) \leq 0, \ \forall i \in J \end{cases}$$

Theorem 8.2.1. *Assume that Problem (P) verifies Slater Condition If (P) has a solution $\overline{x}$, then there exists scalars $(\lambda_i)_{i \in I}, (\mu_i)_{i \in J}$ such that*
(i) $\forall i \in I$, $\lambda_i \geq 0, \lambda_i f_i(\overline{x}) = 0$,
(ii) $\forall i \in J$, $\mu_i \geq 0, \mu_i g_i(\overline{x}) = 0$,
(iii) $0 \in \{\nabla f(\overline{x})\} + \sum_{i \in I} \lambda_i \partial f_i(\overline{x}) + \sum_{i \in J} \mu_i \{\nabla g_j(\overline{x})\}$.

Proof. If $\nabla f(\overline{x}) = 0$, the theorem is true with $\lambda_i = 0$, $\forall i \in I$ and $\mu_i = 0$, $\forall i \in I$.
Assume $\nabla f(\overline{x}) \neq 0$. Let $\alpha = f(\overline{x})$, $\text{lev}_{\leq\alpha}(f) = \{x \in U \mid f(x) \leq \alpha\}$ and $\text{lev}_{<\alpha}(f) = \{x \in U \mid f(x) < \alpha\}$. Obviously, we have $\text{lev}_{<\alpha}(f) \subset \text{i}(\text{lev}_{\leq\alpha}(f))$. We claim that the set $\{x \in U \mid f(x) < \alpha\}$, and hence $\text{i}(\text{lev}_{\leq\alpha}(f))$, is non-empty. Indeed, let $h \in \mathbb{R}^n$ satisfy $\nabla f(\overline{x}) \cdot h < 0$. Since f is differentiable at $\overline{x}$, we have $\forall t > 0$ sufficiently small, $f(\overline{x}+th) - f(\overline{x}) = t(\nabla f(\overline{x}) \cdot h + \theta(t))$ where $\theta(t)$ converges to 0 when t converges to 0. Hence for $t > 0$ small enough, $f(\overline{x} + th) < f(\overline{x}) = \alpha$. Our claim is true.
To end the proof of the theorem, we will proceed by two steps. The first is to prove the result under a stronger assumption and the second step is to construct an equivalent problem that satisfies the assumption of Step 1.

Step 1
First assume that $\nabla f(x) \neq 0$, $\forall x \in U$. We will prove that $\text{i}(\text{lev}_{\leq\alpha}(f)) = \text{lev}_{<\alpha}(f)$. Assume the contrary. There exists $x \in \text{i}(\text{lev}_{\leq\alpha}(f))$ with $f(x) = \alpha$. Then there exists a neighborhood V of x such that $f(y) \leq \alpha = f(x)$, $\forall y \in V$. But this implies $\nabla f(x) = 0$ which is a contradiction with the assumption that $\nabla f(x) \neq 0$, $\forall x \in U$.
Let C be the set of feasible points $\{x \mid f_i(x) \leq 0, \ \forall i \in I$ and $g_i(x) \leq$

$0, \forall i \in J\}$. Then C is nonempty since $\overline{x} \in C$. We have $\mathrm{i}(\mathrm{lev}_{\leq\alpha}(f)) \cap C = \emptyset$. From lemma 8.2.1, there exists $p \in (N_{\mathrm{lev}_{\leq\alpha}(f)}(\overline{x}) \cap -N_C(\overline{x})) \setminus \{0\}$. The Slater condition implies (see Corollaries 7.2.2 and 7.2.3) that

$$N_C(\overline{x}) = \sum_{i \in I(\overline{x})} \{\lambda_i \partial f_i(\overline{x}) \mid \lambda_i \geq 0\} + \sum_{i \in J(\overline{x})} \{\mu_i \nabla g_i(\overline{x}) \mid \mu_i \geq 0\},$$

where $I(\overline{x}) = \{i \in I \mid f_i(\overline{x}) = 0\}, J(\overline{x}) = \{i \in J \mid g_i(\overline{x}) = 0\}$.
From lemma 8.2.2, $N_{\mathrm{lev}_{\leq\alpha}(f)}(\overline{x}) = \{\lambda \nabla f(\overline{x}) \mid \lambda \geq 0\}$. Since $p \in N_{\mathrm{lev}_{\leq\alpha}(f)}(\overline{x}) \cap -N_C(\overline{x})$ and $p \neq 0$, we write $p = \lambda \nabla f(\overline{x}) = -\sum_{i \in I(\overline{x})} \lambda'_i q_i - \sum_{i \in J(\overline{x})} \mu'_i r_i$ where $\lambda > 0, q_i \in \partial f_i(\overline{x}), \lambda'_i \geq 0, \forall i \in I(\overline{x})$, and $r_i = \nabla g_i(\overline{x}), \mu'_i \geq 0, \forall i \in J(\overline{x})$. Define $\lambda_i = \lambda'_i/\lambda, \forall i \in I(\overline{x}), \mu_i = \mu'_i/\lambda, \forall i \in J(\overline{x})$ and $\lambda_i = 0, \forall i \notin I(\overline{x})$, $\mu_i = 0, \forall i \notin J(\overline{x})$. We obtain that

$$0 \in \{\nabla f(\overline{x})\} + \sum_{i \in I} \lambda_i \partial f_i(\overline{x}) + \sum_{i \in J} \mu_i \{\nabla g_i(\overline{x})\},$$

which is Assertion (iii). Moreover, by construction of $(\lambda_i)_{i \in I}, (\mu_i)_{i \in J}$, one obtains:
(i) $\forall i \in I,\ \lambda_i \geq 0,\ \lambda_i f_i(\overline{x}) = 0$,
(ii) $\forall i \in J,\ \mu_i \geq 0,\ \mu_i g_i(\overline{x}) = 0$.

Step 2
Since f is continuously differentiable, there exists an open convex set V containing $\overline{x}$ such that $\nabla f(x) \neq 0, \forall x \in V$. Let $\tilde{f}$ denote the restriction of f to V, note that $\tilde{f}$ is quasi-convex. Then $\overline{x}$ is a solution to Problem $(\tilde{P})$:

$$(\tilde{P}) \qquad \begin{array}{l} \min \tilde{f}(x) \text{ under the constraints} \\ \begin{cases} x \in V \\ f_i(x) \leq 0, \ \forall i \in I \\ g_i(x) \leq 0, \ \forall i \in J \end{cases} \end{array}$$

Let x_0 be the point of U which satisfies Slater Condition. For $\lambda \in]0,1[$ sufficiently small, we have successively: $\lambda x_0 + (1-\lambda)\overline{x} \in V$, $\forall i \in I$, $f_i(\lambda x_0 + (1-\lambda)\overline{x}) < 0$ and $\forall i \in J$, $g_i(\lambda x_0 + (1-\lambda)\overline{x}) \leq 0$.
That means for $\lambda > 0$ small enough, $\lambda x_0 + (1-\lambda)\overline{x}$ verifies Slater Condition for Problem $(\tilde{P})$. Since $\nabla \tilde{f}(x) = \nabla f(x) \neq 0, \forall x \in V$, one can apply the results of Step 1. The proof is now complete. □

Theorem 8.2.2. *If $(\overline{x}, (\lambda_i)_{i \in I}, (\mu_i)_{i \in J})$ satisfy Assertions (i), (ii), (iii) of theorem 8.2.1 and if $\nabla f(\overline{x}) \neq 0$, then $\overline{x}$ is a solution to (P).*

Proof. Let $C = \{x \in U \mid f_i(x) \leq 0, \forall i \in I \text{ and } g_i(x) \leq 0, \forall i \in J\}$. Conditions (i), (ii), (iii) of Theorem 8.2.1 state that $\overline{x} \in C$ and $-\nabla f(\overline{x}) \in N_C(\overline{x})$. Thus we have: $\nabla f(\overline{x}) \cdot (x - \overline{x}) \geq 0, \forall x \in C$. From Lemma 8.2.2, we have $f(x) \geq f(\overline{x})$, $\forall x \in C$. □

Remark 8.2.1. Condition $\nabla f(\overline{x}) \neq 0$ is very important for theorem 8.2.2 as it is shown in the following example. Consider the problem $\min\{f(x) = x^3 \mid x \in \mathbb{R}, x \geq -1\}$. There is a unique solution which is $\overline{x} = -1$. While both $(x = -1, \lambda = 3)$ and $(x = 0, \lambda = 0)$ satisfy Conditions (i), (ii), (iii) of Theorem 8.2.1, only $\overline{x}$ is a solution because $\nabla f(-1) = -3$ while $\nabla f(0) = 0$.

8.3 Differentiable optimization

In this section the function to be minimized is not assumed to be convex or quasi-convex, and the functions which define the constraints are not assumed to be convex. But we assume they are continuously differentiable. The proofs of the results of this section are borrowed from Pallu de la Barrière in his unpublished course at the University of Paris 6 in the 1980's.

Let f be a function from an open, convex, nonempty set U of $\mathbb{R}^n$ into $\mathbb{R}$. We say that f is *locally minimal* at $\overline{x}$ under the constraints $x \in \Gamma$, if $\overline{x} \in \Gamma$ and if there exists a neighborhood V of $\overline{x}$ such that $\forall x \in V \cap \Gamma$, $f(x) \geq f(\overline{x})$.

Consider the following problem (P):

$$(\tilde{P}) \qquad \min f_0(x) \text{ under the constraints } \begin{cases} x \in U \\ f_i(x) \leq 0, \ \forall i \in I \\ g_i(x) = 0, \ \forall i \in K \end{cases}$$

where f_0 and f_i, for $i \in I$, g_i, for i$\in K$ are continuously differentiable functions from an open, convex, nonempty set U of $\mathbb{R}^n$ into $\mathbb{R}$.

Lemma 8.3.1. *Let f be a differentiable function from an open, convex set $U \subset \mathbb{R}^m \times \mathbb{R}^p \times \mathbb{R}^q$ into $\mathbb{R}$. We suppose $0 \in U$. Suppose that the function $f : (x, y, z) \to f(x, y, z)$ is locally minimal at 0 under the constraints $x \geq 0, y = 0$. Then:*

$$f'_x(0,0,0) \geq 0, \ \ f'_z(0,0,0) = 0.$$

Proof. Let $C = \{(x, y, z) \in \mathbb{R}^m \times \mathbb{R}^p \times \mathbb{R}^q \mid x \geq 0, y = 0\}$. Let $(r, s, t) \in C$. Then there exists $\lambda_1 > 0$ such that $\forall \lambda \in [0, \lambda_1[$, one has $(\lambda r, \lambda s, \lambda t) \in C \cap U$. The function F defined by $\forall \lambda \geq 0$, $F(\lambda) = f(\lambda r, \lambda s, \lambda t)$ is locally minimal at 0. Hence, $F'(0) \geq 0$, i.e.

$$f'(0,0,0) \cdot (r, s, t) \geq 0, \ \forall (r, s, t) \in C.$$

In particular, $f'_x(0,0,0) \cdot r \geq 0$, $\forall r \in \mathbb{R}^m$ such that $r \geq 0$, $f'_z(0,0,0) \cdot t \geq 0$, $\forall t \in \mathbb{R}^q$. Hence: $f'_x(0,0,0) \geq 0$ and $f'_z(0,0,0) = 0$. □

Let x be a *feasible point*, i.e. $f_i(x) \leq 0$, $\forall i \in I$, and $g_i(x) = 0$, $\forall i \in K$. Let $I(x) = \{i \mid f_i(x) = 0\}$. We say that the *constraints of Problem (P) are regular at* x if the gradients $(\nabla f_i(x))_{i \in I(x)}$, $(\nabla g_i(x))_{i \in K}$ are linearly independent.

Theorem 8.3.1. *Suppose that f_0 is locally minimal at $\overline{x}$ under the constraints of Problem (P). Suppose also that the constraints are regular at $\overline{x}$. Then there exists non-negative scalars $(\lambda_i)_{i\in I}$, and scalars $(\mu_i)_{i\in K}$ such that:*
(i) $\nabla f_0(\overline{x}) = -\sum_{i\in I} \lambda_i \nabla f_i(\overline{x}) + \sum_{i\in K} \mu_j \nabla g_j(\overline{x})$,
(ii) $\lambda_i f_i(\overline{x}) = 0,\ \forall i \in I$.

Proof. Observe that f_0 is also locally minimal at $\overline{x}$ under the constraints

$$\begin{cases} f_i(x) \leq 0, & \forall i \in I(\overline{x}), \\ g_i(x) = 0, & \forall i \in K \end{cases}$$

where $I(\overline{x}) = \{i \in I \mid f_i(\overline{x}) = 0\}$.
Suppose that $I(\overline{x})$ has cardinal J and K has cardinal r. Since, by assumption, the constraints are regular at $\overline{x}$, there exists $\theta_1, \ldots, \theta_q$, vectors of $\mathbb{R}^n$, with $q = n - J - r$, such that the matrix $((\nabla f_i(\overline{x}))_{i\in I}; (\nabla g_i(\overline{x}))_{i\in K}, \theta_1, \ldots, \theta_q)$ is invertible. Define $\tilde{\theta}_i(x) = \theta_i \cdot (x - \overline{x}),\ \forall i = 1, \ldots, q, \forall x \in \mathbb{R}^n$, and the map φ from $\mathbb{R}^n$ into $\mathbb{R}^n$ by

$$\varphi(x) = ((-f_i(x))_{i\in I(\overline{x})}, (g_j(x))_{i\in K}, (\tilde{\theta}_i(x))_{i=1,\ldots,q}).$$

One has $\varphi(\overline{x}) = 0$, and $J_\varphi(\overline{x})$ is invertible, where $J_\varphi(\overline{x})$ is the Jacobian matrix of φ at $\overline{x}$. From the Local inversion theorem (Theorem A.3.1, in Appendix), φ has, in a neighborhood V of $\overline{x}$, an inverse φ^{-1} which is continuously differentiable.
Define, on $\varphi(V)$, the map $F = f_0 \circ \varphi^{-1}$. One has: $F(u, v, w) = f_0(x)$ if $\varphi(x) = (u, v, w)$. In particular, $F(0, 0, 0) = f_0(\overline{x})$. Hence $F(u, v, w) \geq F(0, 0, 0)$ for $u \geq 0, v = 0$. From Lemma 8.3.1, $F'_u(0, 0, 0) \geq 0$ and $F'_w(0, 0, 0) = 0$, that means: $F'(0, 0, 0) = (\lambda, \mu, 0)$ with $\lambda \in \mathbb{R}^J_+$ and $\mu \in \mathbb{R}^r$. But $\nabla f_0(\overline{x}) = F'(0, 0, 0)(J_\varphi(\overline{x}))$. Since

$$J_\varphi(\overline{x}) = ((-\nabla f_i(\overline{x}))_{i\in I(\overline{x})}, (\nabla g_i(\overline{x}))_{i\in K}, (\theta_i)_{i=1,\ldots,q}),$$

one gets:

$$\nabla f_0(\overline{x}) = -\sum_{i\in I(\overline{x})} \lambda_i \nabla f_i(\overline{x}) + \sum_{i\in K} \mu_i \nabla g_i(\overline{x})$$

with $\lambda_i \geq 0,\ \forall i \in I(\overline{x})$. Define $\lambda_i = 0,\ \forall i \notin I(\overline{x})$. Relation (i) is thus proved. We have $\lambda_i f_i(\overline{x}) = 0, \forall i \in I$, that is Condition (ii). □

Exercises

Exercise 8.1. Let f be a quasi-convex function from an open, convex, non-empty set $U \subset \mathbb{R}^n$. Let $x_1, x_2 \in U$. Show that, if $\forall \lambda \in [0, 1]$, $f(x_1 + \lambda(x_2 - x_1)) \geq f(x_1)$, then the function $g : [0, 1] \to \mathbb{R}$ defined by $g(\lambda) = f(x_1 + \lambda(x_2 - x_1))$ is not decreasing.
Hint: Let $0 < \lambda_1 < \lambda_2$ and $y_1 = x_1 + \lambda_1(x_2 - x_1), y_2 = x_1 + \lambda_2(x_2 - x_1)$. Then $y_1 = x_1 + (\lambda_1/\lambda_2)(y_2 - x_1)$.

Exercise 8.2. Let f be a quasi-convex function from $\mathbb{R}^n$ into $\mathbb{R}$. Let $\overline{x}$ verify $\overline{x} \in V, f(x) \geq f(\overline{x}), \forall x \in V$, where V is a convex set of $\mathbb{R}^n$. We denote by $[a, b]$ the set $\{x \mid x = a + \lambda(b - a), \lambda \in [0, 1]\}$.
(i) Show that, if there exists $y \notin V$ such that $f(y) \leq f(\overline{x})$, then $\forall x \in [\overline{x}, y] \cap V, f(x) = f(\overline{x})$. If for any $y \in V$ there exists z such that $f(z) < f(\overline{x})$ and $y \in [\overline{x}, z]$, then $f(x) = f(\overline{x})$, $\forall x \in V$.
(ii) If moreover: $\forall x \in \mathbb{R}^n, \exists y \in [\overline{x}, x] \cap V$ such that $f(y) > f(\overline{x})$, then $\overline{x}$ is a global minimum of f.

Exercise 8.3. we recall that a function f from $\mathbb{R}^n$ into $\mathbb{R}$ is said to be *strictly quasi-convex* if it is quasi-convex and if $\forall x, \forall y, f(x) < f(y) \Rightarrow \forall \lambda \in]0, 1[, f(\lambda x + (1 - \lambda)y) < f(y)$.
(i) Show that any convex function is strictly quasi-convex.
(ii) Let α be such that the set $\{x \in \mathbb{R}^n \mid f(x) < \alpha\} \neq \emptyset$. Show that $\{x \in \mathbb{R}^n \mid f(x) < \alpha\} = \{x \in \mathbb{R}^n \mid f(x) \leq \alpha\}$.
Hint : Let x satisfy $f(x) = \alpha$ and $x_0 \in \{x \in \mathbb{R}^n \mid f(x) < \alpha\}$. Then we have $x = \lim_{m \to +\infty} x_m$, where $x_m = (1/m)x_0 + (1 - (1/m))x$.
(iii) Show that any local minimum of a strictly quasi-convex function is a global minimum.

Exercise 8.4. Let $U = (\mathbb{R}^2_+ \setminus \{0\}) \times (\mathbb{R}^2_+ \setminus \{0\})$.
(i) Show that the function $f(x, y) = -xy$ is quasi-convex in U.
(ii) Solve

$$\min\{f(x, y) \mid (x, y) \in U, x^2 + y^2 \leq 1\}.$$

Exercise 8.5. Solve

$$\min\{(3\sqrt{2}x + 3y - 1)$$

under the constraints

$$\begin{cases} (x, y) \in \mathbb{R}^2 \\ x^2 + \frac{y^2}{2} = 1\} \end{cases}$$

Exercise 8.6. Consider the ellipsoid (E) defined by the equations

$$\begin{cases} 2x^2 + y^2 - 4 = 0 \\ x + y + z = 0 \end{cases}$$

(i) Find the point of (E) which maximizes the Euclidean distance between (E) and the axis $0y$.
(ii) Find the point of (E) which minimizes the Euclidean distance between (E) and the axis $0y$.

A. Appendix

A.1 Some preliminaries on topology

We consider $\mathbb{R}^n$ with the Euclidean topology. The *norm* of an element $x \in \mathbb{R}^n$ will be denoted by $\|x\|$ and it is equal to $\sqrt{x_1^2 + \ldots + x_n^2}$ where $x = (x_1, \ldots, x_n)$. The *distance* between two points x, y of $\mathbb{R}^n$ is $\|x - y\|$.
We denote by $B(x,r)$ the open ball centered at x with radius r, i.e. $B(x,r) = \{y \in \mathbb{R}^n \mid \|y - x\| < r\}$. The closed ball $\overline{B(x,r)}$ is the set $\{y \in \mathbb{R}^n \mid \|y - x\| \leq r\}$. A subset $U \subset \mathbb{R}^n$ is *open* if for any $x \in U$, there exists $r > 0$ such that $B(x,r) \subset U$. A subset V of $\mathbb{R}^n$ is *closed* if its complement $\mathbb{R}^n \setminus V$ is open. One can check that the open ball $B(x,r)$ is open (in the sense defined above) for any $x \in \mathbb{R}^n$, any $r > 0$. Similarly, the closed ball $\overline{B(x,r)}$ is closed in the sense defined above.
A *mapping* (or a *map*) f from $\mathbb{R}^p$ into $\mathbb{R}^q$ is *continuous* at $a \in \mathbb{R}^p$ if for any open set $V \in \mathbb{R}^q$ containing $f(a)$ there exists an open set $U \in \mathbb{R}^p$ containing a such that $\forall x \in U,\ f(x) \in V$. This mapping is said to be continuous if it is continuous everywhere in $\mathbb{R}^p$. We can define similarly the continuity of a mapping from a set $X \subset \mathbb{R}^p$ into $\mathbb{R}^q$. A mapping f is continuous at a point $a \in X$ if for any open set $V \in \mathbb{R}^q$ containing $f(a)$ there exists an open set $U \in \mathbb{R}^p$ containing a such that $\forall x \in U \cap X,\ f(x) \in V$. One can check that it is equivalent to say that a mapping f from a set $X \subset \mathbb{R}^p$ into $\mathbb{R}^q$ is continuous at a point $a \in X$ if $\forall \varepsilon > 0, \exists \alpha > 0$ such that $\forall x \in X,\ \|x - a\| < \alpha \Rightarrow \|f(x) - f(a)\| < \varepsilon$.
We say that the collection of open sets of $\mathbb{R}^n$, $\{U_i\}_{i \in I}$ is a *covering* of a set $A \subset \mathbb{R}^n$ if $A \subset \cup_{i \in I} U_i$. A set $A \subset \mathbb{R}^n$ is *compact* if from any covering of A there exists a *finite* subcovering of A.

Let us recall the following properties.
A set A of $\mathbb{R}^n$ is compact if, and only if, it is closed and bounded (i.e. there exists $r > 0$ such that $\forall x \in A,\ \|x\| \leq r$).
If f is a continuous mapping from $\mathbb{R}^p$ into $\mathbb{R}^q$, then $f(A)$ is compact in $\mathbb{R}^q$ if A is compact in $\mathbb{R}^p$. If f is a real-valued function defined on $\mathbb{R}^n$ then f has a maximum and a minimum on any compact set of $\mathbb{R}^n$. Moreover, these minimum and maximum are attained.
A point $a \in \mathbb{R}^n$ is a *cluster* point to a set $A \subset \mathbb{R}^n$ if it is the limit of a sequence of points in A. In particular, if A is a sequence of points of $\mathbb{R}^n$, then

a is a cluster point (or an *accumulation* point) of this sequence if there exists a subsequence which converges to a. One can check that a is the limit point of a sequence if, and only if, it is the *unique* cluster point of the sequence. Let $(a_i)_{i\in\mathbb{N}}$ be a *bounded* sequence of $\mathbb{R}^n$. Then $(a_i)_{i\in\mathbb{N}}$ admits a *converging* subsequence.

A.2 The Mean value theorem

In this section, we will give the Mean value theorem for a real-valued function defined on an interval of $\mathbb{R}$. It will be done in two steps. The first is to prove the Rolle's theorem. The Mean value theorem will be deduced from the Rolle's theorem. This theorem is required in Exercise 5.14 and in Chapter 8.

Theorem A.2.1 (Rolle's Theorem). *Let f be a function from $[a,b]$ into $\mathbb{R}$, where $a<b$. We assume that f is continuous on $[a,b]$, differentiable on $]a,b[$ and satisfies $f(a)=f(b)$. Then there exists a point $c\in\,]a,b[$ such that $f'(c)=0$.*

Proof. If f is constant on $[a,b]$ the derivative is equal to zero and the theorem is proved.
Since f is continuous there will be a minimum m and a maximum M of f on $[a,b]$; if f is not constant, either $m<f(a)$ or $M>f(a)$. Without loss of generality, we may assume $m<f(a)$. There exists $c\in[a,b]$ such that $f(c)=m$. Hence $c\neq a, c\neq b$, i.e. $c\in\,]a,b[$.
Consider the ratio $\varphi(h)=(f(c+h)-f(h))/h$, with $h\neq 0$ and h sufficiently small such that $c+h\in[a,b]$. When h tends to 0, $\varphi(h)$ tends to $f'(c)$. We have $f(c+h)-f(h)=f(c+h)-m\geq 0$.
For $h>0$, since $\varphi(h)\geq 0$, we deduce $f'(c)\geq 0$, while for $h<0$, since $\varphi(h)\leq 0$, we deduce $f'(c)\leq 0$. Thus $f'(c)=0$ and the theorem is proved. □

Theorem A.2.2 (Mean value theorem). *Let f be a function from $[a,b]$ into $\mathbb{R}$, where $a<b$. We assume that f is continuous on $[a,b]$ and differentiable on $]a,b[$. There exists a point $c\in\,]a,b[$ such that $f(b)-f(a)=(b-a)f'(c)$.*

Proof. Obviously, we assume that $a\neq b$. Define $A=(f(b)-f(a))/(b-a)$. Consider the function φ defined on $[a,b]$ by $\varphi(x)=f(x)-A(x-a)$. We have $\varphi(a)=\varphi(b)$. Apply Rolle's theorem. There exists $c\in\,]a,b[$ such that $\varphi'(c)=0$. But since $\varphi'(c)=f'(c)-A$, the proof is complete. □

A.3 The Local inversion theorem

Lemma A.3.1 (Contraction theorem). *Let F be a closed, nonempty set of $\mathbb{R}^n$. Let f be a contraction of F, i.e. a mapping from F into F which satisfies:*

$$\exists k \in]0,1[, \forall x \in F, \forall y \in F, \ \|f(x) - f(y)\| \leq k\|x - y\|.$$

Then there exists a unique fixed-point of f, i.e., a unique $\overline{x}$ which satisfies $\overline{x} = f(\overline{x})$. Moreover, $\overline{x} = \lim_{n \to +\infty} f^n(x)$, $\forall x \in F$.

Proof. Let $x \in F$. Let n, m be two integers which satisfy $n > m$. Let $r = n - m$. By induction, we have:

$$\|f^n(x) - f^m(x)\| \leq k^m \|x - f^r(x)\| \tag{A.1}$$

But $\|x - f^r(x)\| \leq \|x - f(x)\| + \|f(x) - f^2(x)\| + \ldots + \|f^{r-1}(x) - f^r(x)\| \leq (1 + k + \ldots + k^{r-1})\|x - f(x)\| \leq (1/(1-k))\|x - f(x)\|$. Hence:

$$\|f^n(x) - f^m(x)\| \leq \frac{k^m}{1-k}\|x - f(x)\|. \tag{A.2}$$

This implies $\|f^n(x)\| \leq (k^m/(1-k))\|x - f(x)\| + \|f^m(x)\|$. Fix m. The sequence $(f^n(x))_{n \in \mathbb{N}}$ is bounded. There exists a subsequence $(f^{n_k}(x))_{k \in \mathbb{N}}$ which converges to an element $\overline{x}$. Inequality (A.2) is true $\forall m, \forall n_k > m$. Hence:

$$\forall m \in \mathbb{N}, \ \|\overline{x} - f^m(x)\| \leq \frac{k^m}{1-k}\|x - f(x)\|. \tag{A.3}$$

Let m converge to $+\infty$. It follows from inequality (A.3) that $f^m(x) \to \overline{x}$ when $m \to +\infty$.
We now show that $\overline{x}$ is a fixed-point of f. Indeed:

$$\|f(\overline{x}) - f^{m+1}(x)\| \leq k\|\overline{x} - f^m(x)\|.$$

Since $f^m(x)$ and $f^{m+1}(x)$ converge to $\overline{x}$, and since $k < 1$, we have $\overline{x} = f(\overline{x})$. Such a $\overline{x}$ is unique, because if there exists another fixed-point $\overline{x}'$, we will have a contradiction:

$$\|\overline{x} - \overline{x}'\| = \|f(\overline{x}) - f(\overline{x}')\| \leq k\|\overline{x} - \overline{x}'\| < \|\overline{x} - \overline{x}'\|.$$

□

Let $f : D \to \mathbb{R}^p$ be a function and $x \in \mathrm{i}(D)$ where $D \subset \mathbb{R}^n$. We recall that the function f is *differentiable* at x if there exists a linear map $\varphi : \mathbb{R}^n \to \mathbb{R}^p$ (such a φ will be unique) such that $\lim_{z \to x}(f(z) - f(x) - \varphi(z - x))/\|z - x\| = 0$. Such a φ is called *differential* of f at x and noted $df(x)$ or f'.
Assume that f is differentiable at x. Then, for all $y \neq 0$, one has

$$\lim_{\lambda > 0, \lambda \to 0} \frac{f(x + \lambda y) - f(x) - df(x)(\lambda y)}{\lambda \|y\|} = 0.$$

In particular, for $i = 1, \ldots, n$, for $j = 1, \ldots, p$,

$$df_j(x) \cdot e^i = \lim_{\lambda \to 0} \frac{f_j(x + \lambda e^i) - f_j(x)}{\lambda} = \left(\frac{\partial f_j}{\partial x_i}\right)(x),$$

where e^i is the unit vector of the basis of $\mathbb{R}^n$.
Hence if $y = (y_1, \ldots, y_n)$, then $(df(x)(y)) = (\sum_{i=1}^n y_i(\partial f_j/\partial x_i)(x))_{j=1,\ldots,p}$. Consequently, the differential $df(a)$ of f at a may be identified to the Jacobian matrix $J_f(a) = ((\partial f_i/\partial x_j)(a))_{\substack{i=1,\ldots,n \\ j=1,\ldots,p}}$ where $f_i(a)$ is the i-th component of $f(a)$.

Theorem A.3.1 (Local inversion theorem). *Let f be a continuously differentiable mapping from an open, nonempty convex set U of $\mathbb{R}^n$ into $\mathbb{R}^n$. Let $a \in U$ verify $f(a) = 0$. Assume that the Jacobian matrix $J_f(a)$ is invertible. Then there exists an open set $V \subset U$, which contains a, and an open set W containing 0 such that f is one-to-one from V into W.*
Moreover, the inverse f^{-1} is continuously differentiable on W.

Proof. Define $\forall x \in U$, $h(x) = (df(a))^{-1}f(x)$ and $g(x) = x - h(x)$. We will prove that h is one-to-one "on a neighborhood of 0", and that both h and its inverse are continuous and differentiable.
We have: $g(a) = a$, $dh(a) = I$ and $dg(a) = 0$, where I is the Identity mapping of $\mathbb{R}^n$. Hence there exists $r > 0$ such that $(df(x))^{-1}$ exists and:

$$\|x - a\| \le 2r \Rightarrow \|dg(x)\| \le \frac{1}{2}. \tag{A.4}$$

Since

$$g(x') - g(x) = \int_0^1 dg(x + \lambda(x' - x))(x' - x)d\lambda,$$

from Inequality (A.4) one gets:

$$\forall x \in \overline{B(a, 2r)}, \forall x' \in \overline{B(a, 2r)},\ \|g(x) - g(x')\| \le \frac{1}{2}\|x - x'\| \tag{A.5}$$

Hence:

$$\|x - a\| \le 2r \Rightarrow \|g(x) - a\| \le \frac{1}{2}\|x - a\| \tag{A.6}$$

We see that g maps $\overline{B(a, r)}$ into $\overline{B(a, r/2)}$.
We will show that h is one-to-one from $h^{-1}(\overline{B(0, r/2)})$ into $\overline{B(0, r/2)}$. Take $y \in \overline{B(0, r/2)}$ and define

$$\phi_y(x) = y + g(x) = y + x - h(x)$$

From Relation (A.6) one gets:

$$\|x - a\| \le r, \|y\| \le \frac{r}{2} \Rightarrow \|\phi_y(x) - a\| \le r.$$

This relation shows that ϕ_y maps $\overline{B(a, r)}$ into $\overline{B(a, r)}$. From Relation (A.5), we have:

$$\forall x \in \overline{B(a, r)}, \forall x' \in \overline{B(a, r)},\ \|\phi_y(x) - \phi_y(x')\| = \|g(x) - g(x')\| \le \frac{1}{2}\|x - x'\|.$$

Therefore, ϕ_y is a contraction from $\overline{B(a,r)}$ into $\overline{B(a,r)}$. From lemma A.3.1, ϕ_y has a unique fixed-point which by definition of ϕ_y is precisely the unique solution to the equation $h(x) = y$. We also have $h^{-1}(\overline{B(0,r/2)}) \subset \overline{B(a,r)}$. Denote by h^{-1} the inverse mapping of h from $\overline{B(0,r/2)}$ into $h^{-1}(\overline{B(0,r/2)})$. We claim that h^{-1} is continuous. For that, let us consider y, y' in $\overline{B(0,r/2)}$. Let $x = h^{-1}(y)$ and $x' = h^{-1}(y')$. Since by definition, $x = g(x) + h(x)$, we see that $\|x - x'\| \leq \|g(x) - g(x')\| + \|h(x) - h(x')\|$. Noticing that x and x' are in $\overline{B(a,r)}$, it follows from Condition (A.5):

$$\|x - x'\| \leq \frac{1}{2}\|x - x'\| + \|h(x) - h(x')\|$$

and hence:

$$\|h^{-1}(y) - h^{-1}(y')\| \leq 2\|y - y'\| \tag{A.7}$$

which shows that h^{-1} is continuous.
Now, we want to show that h^{-1} is differentiable in $B(0,r/2)$. Fix some $\overline{y} \in B(0,r/2)$, and define $\overline{x} = h^{-1}(\overline{y})$ and $A = (dh(\overline{x}))^{-1}$. For all $y \in B(0,(r/2))$, we let $x = h^{-1}(y)$. It follows:

$$\|h^{-1}(y) - h^{-1}(\overline{y}) - A(y - y_1)\| = \|x - \overline{x} - A(h(x) - h(\overline{x}))\| \tag{A.8}$$

Since h is differentiable at point $\overline{x}$, we can write

$$h(x) = h(\overline{x}) + dh(\overline{x})(x - \overline{x}) + \|x - \overline{x}\|\theta(x - \overline{x}),$$

where $\lim \theta(x - \overline{x}) \to 0$ when $x \to \overline{x}$. Relation (A.8) becomes:

$$\|h^{-1}(y) - h^{-1}(\overline{y}) - A(y - \overline{y})\| = \|A(\|x - \overline{x}\|\theta(x - \overline{x}))\| \tag{A.9}$$

From Relation (A.7), $\|x - \overline{x}\| \leq 2\|y - \overline{y}\|$. Hence

$$\frac{\|A(\|x - \overline{x}\|\theta(x - \overline{x}))\|}{\|y - \overline{y}\|} \leq 2\frac{\|A(\|x - \overline{x}\|\theta(x - \overline{x}))\|}{\|x - \overline{x}\|} = 2\|A(\theta(x - \overline{x}))\|.$$

The last member of the previous inequality converges to 0 when y converges to $\overline{y}$. From Relation (A.9), one concludes that h^{-1} is differentiable at $\overline{y}$ and its differential at this point is given by the formula $d(h^{-1})(\overline{y}) = (dh(h^{-1}(\overline{y})))^{-1}$. Since $h^{-1}, dh, (dh)^{-1}$ are continuous, $d(h^{-1})$ is also continuous at $\overline{y}$.
Now, since $f = (df(a))h$, then f has an inverse f^{-1} from $W = B(0,r/2)$ into $V = f^{-1}(W)$ and f^{-1} is continuously differentiable in W. □

A.4 The implicit functions theorem

From the Local inversion theorem, one can get the Implicit Functions Theorem. It is not used in this book. However it may help the readers in their research.

Theorem A.4.1 (The Implicit Functions Theorem). *Let f be a continuously differentiable function from $U \times V$ into $\mathbb{R}^n$ where U is an open, nonempty set of $\mathbb{R}^m$ and V is an open, nonempty set of $\mathbb{R}^n$. Suppose that $f(a,b) = 0$ and that $(J_f)_2 f(a,b)$ is invertible, where $(J_f)_2 f(a,b)$ is the partial derivative of f, with respect to the second variable, at point (a,b). Then there exists an open set $U_1 \subset U$ containing a, an open set $V_1 \subset V$ containing b, an open set W containing 0 and a function $g : U_1 \times W \to V_1$ such that:*

$$\forall x \in U_1, \forall y \in V_1, \forall z \in W, \;\; f(x,y) = z \Leftrightarrow y = g(x,z).$$

Moreover, g is continuously differentiable on $U_1 \times W$. In particular,

$$(x,y) \in U_1 \times W, \;\; f(x,y) = 0 \Leftrightarrow y = \phi(x),$$

where ϕ is continuously differentiable on U_1.

Proof. Define $h : U \times V \to \mathbb{R}^m \times \mathbb{R}^n$ by $h(x,y) = (x - a, f(x,y))$. We have $h(a,b) = (0,0)$ and

$$J_h(a,b) = \begin{pmatrix} I & 0 \\ (J_f)_1 f(a,b) & (J_f)_2 f(a,b) \end{pmatrix} \tag{A.10}$$

The matrix $J_h(a,b)$ is invertible since $(J_f)_2 f(a,b)$ is. Apply the Local inversion theorem. There exists an open set $U_1 \subset U$ containing a, an open set $V_1 \subset V$ containing b, an open set $U_2 \subset \mathbb{R}^m$ containing 0, an open set $W \subset \mathbb{R}^m$ containing 0, such that h is one-to-one from $U_1 \times V_1$ into $U_2 \times W$, and its inverse h^{-1} is continuously differentiable on $U_2 \times W$.
In other words, relation $(x - a, f(x,y)) = (s,z)$ with $(x,y) \in U_1 \times V_1$ and $(s,z) \in U_2 \times W$ is equivalent to $x = a + s$ and $y = g_1(s,z)$ where g_1 is continuously differentiable in $U_2 \times W$. Define $g(x,z) = g_1(x - a, z)$. Then $f(x,y) = z \Leftrightarrow y = g(x,z) = \phi(x)$.
In particular, let ϕ be defined by $\phi(x) = g(x,0)$. then, $f(x,y) = 0 \Leftrightarrow y = g(x,0) = \phi(x)$. The proof is complete. □

Extensions of the Mean value theorem, the local inversion theorem and the Implicit functions theorem, for mappings from a Banach space into a Banach space can be found, e.g. in Cartan [13].

Bibliography

[1] C.D. Aliprantis and K.C. Border. *Infinite Dimension Analysis. A Hitchhiker's Guide.* 2d. ed. New York Heidelberg Berlin: Springer-Verlag, 1999.

[2] K. Arrow, L. Hurwicz, and Editors H. Uzawa. *Studies in Linear and Non-Linear Programming.* Stanford University Press, 1958.

[3] J.P. Aubin. *Mathematical Methods of Game and Economic Theory.* North-Holland Publishing Company, 1979.

[4] A. Balakrishnan, editor. *Control Theory and the Calculus of Variations.* New york: Academic Press, 1969.

[5] L.M. Benveniste and J.A. Scheinkman. On the differentiability of the value function in dynamic models of economics. *Econometrica* **47**, pages 727–732, 1979.

[6] C. Berge. *Espaces Topologiques.* Paris: Dunod, English translation: Topological Spaces, Edinburgh London: Oliver and Boyd, 1963, 1959.

[7] C. Berge and A. Ghouila-Houri. *Programmes, Jeux et Réseaux de Transport.* Paris: Dunod, 1962.

[8] J.F. Bonnans and A. Shapiro. *Perturbation Analysis of Optimization Problems.* New York Heidelberg Berlin: Springer-Verlag, 2000.

[9] K.C. Border. *Fixed point theorems with applications to economics and game theory.* New York: Cambridge University Press, 1985.

[10] J.M. Borwein and A.S. Lewis. *Convex Analysis and Nonlinear Optimization, Theory and Examples.* Canadian Mathematical Society, New York Heidelberg Berlin: Springer-Verlag, 2000.

[11] N. Bourbaki. *Espaces Vectoriels Topologiques, 2ème edition.* Paris: Hermann, 1966.

[12] G. Cagnac, E. Ramis, and J. Commeau. *Nouveau Cours de Mathématiques Spéciales.* Masson & Cie, 1966.

[13] H. Cartan. *Calcul Différentiel.* Hermann Paris, Collection Méthodes, 1967.

[14] F.H. Clarke. *Optimization and Nonsmooth Analysis.* John Wiley & Sons, 1983.

[15] R.A. Dana, C. Le Van, and F. Magnien. On the different notions of arbitrage and existence of equilibrium. *Journal of Economic Theory* **87**, pages 169–193, 1999.

[16] G.B. Dantzig. *Maximization of a linear function of variables subject to linear constraints*, pages 339–347. In Koopmans [41], 1951.

[17] G.B. Dantzig. *Linear Programming and Extensions.* Princeton: Princeton University Press, 1963.

[18] G.B. Dantzig and B.C. Eaves, editors. *Studies in Optimization.* The Mathematical Association of America, 1974.

[19] G.B. Dantzig, A. Orden, and P. Wolfe. Generalized simplex method for minimizing a linear form under linear inequality constraints. *Pacific J. Math.* **5**, pages 183–195, 1955.

[20] G. Debreu. *Theory of Value.* New York: John Wiley, 1959.

[21] W. Fenchel. *Convex Cones, Sets and Functions.* Notes of Lectures at Princeton University, 1951.

[22] J. Gaddum. A theorem on convex cones with applications to linear inequalities. *Proc. Am. Math. Soc.* **3**, pages 37–49, 1950.

[23] D. Gale. *Convex polyhedral cones and linear inequalities*, pages 287–297. In Koopmans [41], 1951.

[24] D. Gale. *The Theory of Linear Economic Models.* New York Toronto London: McGraw-Hill Book Company, 1960.

[25] M. Gerstenhaber. *Theory of convex polyhedral cones*, pages 298–316. In Koopmans [41], 1951.

[26] J.R. Giles. *Convex analysis with applications in differentiation of convex functions.* Boston London Melbourne: Pitman Advanced Publishing Program, 1982.

[27] A.J. Goldman. *Resolution and separation theorems for polyhedral convex sets*, pages 41–51. In Kuhn and Tucker [46], 1956.

[28] A.J. Goldman and A.W. Tucker. *Polyhedral convex cones*, pages 19–51. In Kuhn and Tucker [46], 1956.

[29] A.J. Goldman and A.W. Tucker. *Theory of linear programming*, pages 53–97. In Kuhn and Tucker [46], 1956.

[30] J.-B. Hiriart-Urruty and C. Lemarechal. *Convex Analysis and Minimization Algorithms.* New York Heidelberg Berlin: Springer-Verlag, 1993.

[31] R.B. Holmes. *Geometric Functional Analysis and its Applications.* New York Heidelberg Berlin: Springer-Verlag, 1975.

[32] T. Ichiishi. *Game Theory for Economic Analysis.* New York London: Academic Press, 1983.

[33] V.L. Klee. Convex sets in linear spaces. *Duke. Math. J.* **18**, pages 443–466, 1951.

[34] V.L. Klee. Convex sets in linear spaces, II. *Duke. Math. J.* **18**, pages 875–883, 1951.

[35] V.L. Klee. Convex sets in linear spaces, III. *Duke. Math. J.* **20**, pages 105–112, 1951.

[36] V.L. Klee. Some characterizations of convex polyhedra. *Acta Math.* **102**, pages 79–107, 1951.

[37] V.L. Klee. Strict separation of convex sets. *Proc. Amer. Math. Soc.* **7**, pages 735–737, 1956.

[38] V.L. Klee. Extremal structure of convex sets. *Arch. Math.* **8**, pages 234–250, 1957.

[39] V.L. Klee. Extremal structure of convex sets, II. *Math. Zeit.* **69**, pages 90–104, 1958.

[40] V.L. Klee. *Separation and support properties of convex sets - a survey*, pages 235–304. In Balakrishnan [4], 1969.

[41] T.C. Koopmans, editor. *Activity Analysis of Production and Allocation.* Cowles Commission Monograph No.13, New York: John Wiley and Sons, 1951.

[42] M. Krein and D. Milman. On the extreme points of regularly convex sets. *Studia Math.* **9**, pages 133–138, 1940.

[43] H.W. Kuhn. Solvability and consistency for linear equations and inequalities. *Amer. Math. Monthly.* **63**, pages 217–232, 1956.

[44] H.W. Kuhn and A.W. Tucker, editors. *Contributions to the Theory of Games, Vol. I.* Annals of Mathematics Studies **24**, Princeton: Princeton University Press, 1950.

[45] H.W. Kuhn and A.W. Tucker, editors. *Contributions to the Theory of Games, Vol. II.* Annals of Mathematics Studies **28**, Princeton: Princeton University Press, 1953.

[46] H.W. Kuhn and A.W. Tucker, editors. *Linear Inequalities and Related Systems.* Annals of Mathematics Studies **38**, Princeton: Princeton University Press, 1956.

[47] H. Nikaido. *Convex Structures and Economic Theory.* New York and London: Academic Press, 1972.

[48] R. Pallu de La Barrière. *Unpublished Course.* University of Paris 6, 1980.

[49] R.R. Phelps. *Convex Functions, Monotone Operators and Differentiability.* 2d. ed. New York Heidelberg Berlin: Springer-Verlag, 1993.

[50] A.W. Roberts and D.E. Varberg. *Convex Functions.* New York San Francisco London: Academic Press, 1973.

[51] R.T. Rockafellar. *Convex Analysis.* Princeton: Princeton University Press, 1970.

[52] T. Rockafellar. A combinatorial algorithm for linear programs in the general form. *SIAM J. Appl. Math.* **12**, pages 215–225, 1964.

[53] W. Rudin. *Functional Analysis, second edition.* McGraw-Hill, Inc., 1991.

[54] E. Stiemke. Uber positive losungen homogener linearer gleichungen. *Math. Ann. 76*, pages 340–342, 1915.

[55] A.W. Tucker. Extensions of theorems of Farkas and Stiemke. *Bull. Amer. Math. Soc.* **56**, page 57, 1950.

[56] A.W. Tucker. *Dual systems of homogeneous linear relations*, pages 3–18. In Kuhn and Tucker [46], 1956.

[57] A.W. Tucker. *Combinatorial algebra of linear programs.* In Dantzig and Eaves [18], 1974.

[58] H. Tuy. Convex inequalities and the Hahn-Banach theorem. *Diss. Math.* **47**, 1972.

[59] F. Valentine. *Convex Sets.* New York: McGraw-Hill, 1964.

[60] J. von Neumann. Discussion of a maximum problem. *In Collected Works, Vol. VI, edited by A.H. Taub, New York: Pergamon Press*, 1963.

[61] J. von Neumann and O. Morgenstern. *Theory of Games and Economic Behavior.* Princeton, 1944 (3rd edition 1953).

[62] H. Weyl. Elementare theorie der konvexen polyeder. *Comm. Math. Helv.* **7**, 1935. Also (in English translation by H.W. Kuhn) The elementary theory of convex polyhedra, pages 3–18, in [44].

Index

Printing: Weihert-Druck GmbH, Darmstadt
Binding: Buchbinderei Schäffer, Grünstadt